EURO Advanced Tutorials on Operational Research

Series Editors

M. Grazia Speranza, Brescia, Italy

José Fernando Oliveira, Porto, Portugal

The EURO Advanced Tutorials on Operational Research are a series of short books devoted to an advanced topic—a topic that is not treated in depth in available textbooks. The series covers comprehensively all aspects of Operations Research. The scope of a Tutorial is to provide an understanding of an advanced topic to young researchers, such as Ph.D. students or Post-docs, but also to senior researchers and practitioners. Tutorials may be used as textbooks in graduate courses.

More information about this series at http://www.springer.com/series/13840

Miguel F. Anjos • Manuel V. C. Vieira

Facility Layout

Mathematical Optimization Techniques and Engineering Applications

 Springer

Miguel F. Anjos
School of Mathematics
University of Edinburgh
Edinburgh, UK

Manuel V. C. Vieira
NOVA School of Science and Technology
Universidade NOVA de Lisboa
Caparica, Portugal

ISSN 2364-687X   ISSN 2364-6888   (electronic)
EURO Advanced Tutorials on Operational Research
ISBN 978-3-030-70989-1   ISBN 978-3-030-70990-7   (eBook)
https://doi.org/10.1007/978-3-030-70990-7

This Springer imprint is published by the registered company Springer Nature Switzerland AG.
The registered company address is: Gewerbestrasse 11, 6330 Cham, Switzerland

Preface

Facility layout problems are a general class of operations research problems concerned with finding the optimal physical arrangement of the space inside a facility. Facility layout has been extensively studied since the 1960s. Originally motivated by the physical organization of manufacturing systems, it remains particularly relevant for flexible manufacturing systems that produce an array of different parts because the layout of the production components has a significant impact on the costs and the productivity of these systems. This is nowadays only one of the numerous areas of application for facility layout problems, as evidenced by the large literature in the area.

Facility layout is fundamentally concerned with imposing structure on a given "space", and therefore we have structured the main part of this book according to the properties of the space under consideration and the type of structure to be imposed. We first consider facilities that have a one-dimensional structure or, equivalently, a row-based structure. For example, the rows may be corridors in a building, and the departments will then be the rooms to locate along the corridor in a way that allows efficient access. Chapter 2 considers facilities with a single row, and Chap. 3 considers two or more rows. Chapter 4 considers a single floor in a building, and how to allocate the corresponding space to the departments to be located there. This is generally known in the literature as the unequal-areas facility layout problem, and it can be viewed as a two-dimensional layout problem. Chapter 5 presents extensions of facility layout as well as some related problems. Section 5.1 introduces the quadratic assignment problem, a well-known special case. Section 5.2 concerns re-layout problems, which occur when changes must be made to an existing layout by moving a certain number of departments to new locations. Section 5.3 discusses multi-floor layout problems, which seek the optimal arrangement of departments in a facility with multiple floors. Section 5.4 looks at dynamic versions of layout problems, that is, where the layouts are designed to be optimal over several periods of time, with the requirements varying between time periods. Finally, Chap. 6 illustrates a wide range of applications of facility layout in engineering via a variety of examples from the literature.

We focus on mathematical optimization-based approaches for facility layout, including both linear and nonlinear models. We assume that the reader is familiar with the basics of building such models, as presented, for example, by Williams (2013). The classic text by Horn and Johnson (1990) provides essential background for the more advanced topics discussed in the Appendix.

It is important to realize that nearly all variants of facility layout are NP-hard, so solving these problems to global optimality in a reasonable time is generally difficult. Therefore, in addition to presenting exact methods, we consider how mathematical optimization can be used to obtain approximate solutions, via either convex relaxations or optimization-based heuristics. Beyond the approaches that we present, there is a rich literature on heuristic algorithms that do not apply mathematical optimization. We refer the reader to the survey papers of Meller and Gau (1996), Singh and Sharma (2006), Kothari and Ghosh (2012), and more recently Keller and Buscher (2015) and Anjos and Vieira (2017).

With the exception of Chap. 1, we have placed all the relevant references to the literature in the last section of each chapter. This makes it possible for the interested reader to pursue the ideas presented in greater depth while not interrupting the flow of ideas in the text.

We thank Grazia Speranza and José Fernando Oliveira for the opportunity to contribute this book in the *EURO Advanced Tutorials on Operational Research* series.

Facility layout is an exciting and thriving area of operations research. We hope that this book will make it better known and convince others to explore these surprisingly difficult and challenging problems!

Edinburgh, UK   Miguel F. Anjos
Caparica, Portugal   Manuel V. C. Vieira
November 2020

References

Anjos MF, Vieira MVC (2017) Mathematical optimization approaches for facility layout problems: The state-of-the-art and future research directions. Eur J Oper Res 261(1):1–16

Horn R, Johnson C (1990) Matrix analysis. Cambridge University Press, Cambridge

Keller B, Buscher U (2015) Single row layout models. Eur J Oper Res 245(3):629–644

Kothari R, Ghosh D (2012) The single row facility layout problem: State of the art. Opsearch 49(4):442–462

Meller RD, Gau KY (1996) The facility layout problem: Recent and emerging trends and perspectives. J Manuf Syst 15(5):351–366

Singh SP, Sharma RR (2006) A review of different approaches to the facility layout problems. Int J Adv Manuf Technol 30(5–6):425–433

Williams HP (2013) Model building in mathematical programming. Wiley

Contents

Chapter 1
Motivation

A facility, in the general sense, is a building where people come together for a specific purpose (Heragu 2008). Activities taking place in a facility normally also require the presence and use of various items of equipment (furniture, machines, materials, etc.). A well-managed facility is a means to achieve higher productivity. This motivates interest in identifying and implementing optimal decisions for locating, designing, and building facilities.

This book is concerned with mathematical optimization approaches for *facility layout*, which is one of the fundamental aspects of facility design. Facility layout refers to the physical arrangement of the space inside the facility. A facility may be part or all of a building, an outdoor space, a warehouse, an office building, etc. To apply an optimization approach, we begin by defining a number of *departments*. Examples of departments are offices, cafeterias, classrooms, and manufacturing workstations. Departments generally contain equipment; we consider the need for floorspace to be "equipment," so departments include functional spaces such as corridors and elevators.

According to its function and the equipment it contains, each department has a specified area requirement (the space needed for the equipment to be placed and used effectively), and the area has a shape requirement (or limits on the dimensions). While departments such as cafeterias and offices can be shaped flexibly provided the total area requirement is met, corridors and elevators have fairly fixed dimensions.

Once we have a list of departments with the corresponding area and shape requirements, the facility layout problem is to assign each department to a specific location and, in the case of a flexible department, to determine its shape. Note that the facility itself will have a fixed total area and generally fixed dimensions, and the assignment must respect these requirements as well. The typical objective is to minimize the total distance travelled by people and materials between departments, in order to maximize the efficiency of the operations in the facility. This distance can be calculated using information about the range and frequency of the activities that will be carried out: Simmons (1969) argues that from a functional point

© Springer Nature Switzerland AG 2021
M. F. Anjos, M. V. C. Vieira, *Facility Layout*, EURO Advanced Tutorials
on Operational Research, https://doi.org/10.1007/978-3-030-70990-7_1

of view, it is possible to determine an "optimal solution" based solely on the walking distances and traffic densities. Other possible objectives include reducing congestion, facilitating communication, and increasing safety (Heragu 2008).

An assignment is said to be *optimal* if it provides a total distance travelled that is no greater than the total distance travelled for the other *feasible* assignments, i.e., assignments that satisfy all the requirements of area and shape. The purpose of a facility layout algorithm is to compute such an optimal layout for a given instance of the problem or, alternatively, to provide a competitive layout, ideally with a guarantee that it is close to optimality.

In this book we limit our attention to approaches based on mathematical optimization. We consider both exact methods and heuristics (sometimes called matheuristics) that make use of mathematical optimization approximations and/or relaxations. Specifically, we consider models based on *mixed-integer linear optimization (MILO)*, often referred to as mixed-integer programming or MIP, *semidefinite optimization (SDO)*, also called semidefinite programming or SDP, and *nonlinear optimization*, also called nonlinear programming or NLP. We assume that the reader is familiar with the basics of building mathematical optimization models, as presented, for example, in Williams (2013).

The use of mathematical optimization models for layout dates back at least to the seminal paper of Koopmans and Beckmann (1957), which introduced the famous quadratic assignment problem (see Sect. 5.1). Research into the application of these models to layout problems has thus been carried out for more than 60 years and continues on both the theoretical and applied fronts.

References

Heragu SS (2008) Facilities design. CRC Press
Koopmans TC, Beckmann M (1957) Assignment problems and the location of economic activities. Econometrica 25(1):53–76
Simmons DM (1969) One-dimensional space-allocation algorithm: An ordering algorithm. Operations Research 17(5):812–826
Williams HP (2013) Model building in mathematical programming. Wiley

Chapter 2
Layout on a Single Row

Single-row layout has a simple structure. Given a set of departments to place alongside each other in a single row, the problem typically boils down to finding a permutation of the departments that minimizes the total distance travelled.

2.1 Introductory Example

We start with an example from a manufacturing context. Consider a factory that needs to place 12 machines, described in Table 2.1, in a line. This could be because they are to be placed along a single wall of a large factory, for example. There is traffic (of people and/or parts) between the machines, and Table 2.2 gives the amount of traffic between each pair of machines when the factory is in operation. Given this information, we seek to arrange the machines in a single row so that the total distance travelled when they are operating is minimized. This is an example of the single-row facility layout problem (SRFLP).

Figure 2.1 illustrates the SRFLP in the context of placing the machines along the path of an automated guided vehicle (AGV) transporting material between them; here, the objective is to minimize the distance travelled by the AGV.

To build a mathematical optimization model, we represent each machine as a rectangle of the same length as the machine. The heights of the rectangles can be all equal, say one, because they are not relevant to the optimization. We measure the distances in one dimension, say along the x-axis, and we introduce variables $x_1, x_2, \ldots, x_{12}$ to denote the location along the x-axis of the centre of each machine. Using these variables, we can express the distance between each pair of machines i, j as the absolute value of the difference of the locations of their centres: $|x_i - x_j|$.

© Springer Nature Switzerland AG 2021
M. F. Anjos, M. V. C. Vieira, *Facility Layout*, EURO Advanced Tutorials
on Operational Research, https://doi.org/10.1007/978-3-030-70990-7_2

Table 2.1 Machines in factory

Machine	Type	Length
1	Lathe	15
2	Drill	10
3	Mill	20
4	Punch press	15
5	Centreless grinder	15
6	Shaper	20
7	Planer	20
8	Drill	10
9	Mill	20
10	CNC machine	25
11	Punch press	15
12	Vertical turret lathe	10

Table 2.2 Amount of traffic between machines

Machine	1	2	3	4	5	6	7	8	9	10	11	12
1	–	4	1	4	4	2	0	4	0	0	3	2
2		–	2	4	1	2	0	1	0	0	0	0
3			–	3	4	4	3	1	1	4	1	0
4				–	3	3	1	1	0	0	0	1
5					–	0	0	3	2	0	3	2
6						–	1	1	1	4	1	0
7							–	1	3	2	0	0
8								–	0	0	3	4
9									–	2	3	2
10										–	0	0
11											–	1
12												–

Fig. 2.1 Single-row facility layout with an AGV (automated guided vehicle)

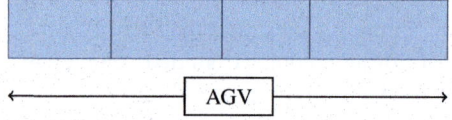

We further denote by c_{ij} the amount of traffic between machines i and j, given in Table 2.2. We can now express the total distance travelled between all the pairs of machines as

$$\sum_{i=1}^{11} \sum_{j=i+1}^{12} c_{ij}|x_i - x_j| = c_{1,2}|x_1 - x_2| + c_{1,3}|x_1 - x_3| + \ldots + c_{11,12}|x_{11} - x_{12}| \quad (2.1)$$

$$= 4 \times |x_1 - x_2| + 1 \times |x_1 - x_3| + \ldots + 1 \times |x_{11} - x_{12}|.$$

Fig. 2.2 Minimum distance to prevent machines i and j from overlapping

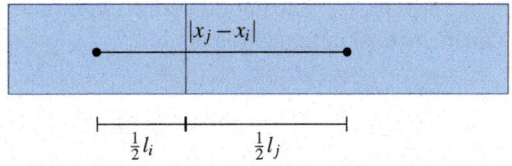

Hence, in mathematical optimization terms, we wish to minimize the objective function (2.1) over all possible values of the x_i variables.

It is straightforward to observe that simply minimizing (2.1) gives an optimal solution in which all the x_i variables are set to the same value (any value). This corresponds to a solution in which the machines are placed on top of each other at the same position, and the total distance travelled is equal to zero.

To prevent machine overlap, we require the distance between each pair of machines i and j to be at least half of the sum of their lengths. This is illustrated in Fig. 2.2. We express this mathematically as

$$|x_i - x_j| \geq \frac{1}{2}(\ell_i + \ell_j), \tag{2.2}$$

where ℓ_i denotes the length of machine i, given in Table 2.1.

Note that constraint (2.2) does not allow any clearance (empty space) between machines i and j. In practice, a given clearance requirement is associated with a machine, and therefore, we can assume that the clearance is included in the given length of the machine. Most other general clearance requirements can also be integrated without difficulty.

When building mathematical optimization models, we must ensure that all the variables are bounded. This consideration leads us to observe that for this example, a row of length $L = \sum_{i=1}^{12} \ell_i = 195$ suffices to find the optimal placement of the machines. This is because the traffic values in Table 2.2 are all non-negative, and therefore, it makes no sense to leave unused space between two machines when minimizing (2.1). In other words, because the weights of the objective function (2.1) are all greater than or equal to zero, the optimization will ensure that the machines are placed next to each other along the x-axis with no unused space between adjacent machines. We can also require the variables x_i to be non-negative without invalidating the formulation. Hence, we can restrict each x_i to have a value between 0 and L: $0 \leq x_i \leq L$.

Beyond ensuring that all the variables are bounded below and above, we wish to ensure that the variables are bounded as tightly as possible. This is helpful because tighter bounds typically reduce the computational time required to determine the optimal solution. We therefore ask now whether $0 \leq x_i \leq L$ can be tightened. Looking again at Fig. 2.2, we observe that because x_i is at the centre of rectangle i, its leftmost possible position in the interval $[0, L]$ is $\frac{1}{2}\ell_i$. Similarly, its rightmost possible position is $L - \frac{1}{2}\ell_i$. This reasoning allows us to tighten the bounds to $\frac{1}{2}\ell_i \leq x_i \leq L - \frac{1}{2}\ell_i$.

We now have all the ingredients to express our machine placement problem using a mathematical optimization model:

$$\text{minimize} \quad \sum_{i=1}^{11} \sum_{j=i+1}^{12} c_{ij} |x_i - x_j|, \tag{2.3}$$

$$\text{subject to} \quad |x_i - x_j| \geq \frac{1}{2}(\ell_i + \ell_j), \quad i, j = 1, \ldots, 12, \ i < j, \tag{2.4}$$

$$\frac{1}{2}\ell_i \leq x_i \leq L - \frac{1}{2}\ell_i, \quad i = 1, \ldots, 12. \tag{2.5}$$

We will sometimes use the abbreviation "min" and write below it the variables involved in the optimization.

2.1.1 On Convexity and Linearity

Even though the mathematical optimization problem (2.3–2.5) is a correct model for the example, there are better formulations in practice. We advocate the use whenever possible of *convex optimization formulations* because they have two important advantages. First, for a convex optimization problem (defined below), any local optimal solution is also a global optimal solution. Second, there exist polynomial-time interior-point algorithms for convex optimization that have been shown to work well in practice. An algorithm is said to run in *polynomial time* if the number of steps required to complete the algorithm for a given input is bounded above by a polynomial function (of fixed degree) of the size of the input. Polynomial-time algorithms are generally considered to be efficient.

A convex optimization problem consists of minimizing a convex objective function (or maximizing a concave function) over a convex set of possible solutions. These concepts are defined as follows.

Definition 2.1 (See Fig. 2.3.) A function $f : \mathfrak{R}^n \to \mathfrak{R}$ is *convex* if its domain is convex and if for all $\overline{x}, \overline{y}$ in its domain and $\tau \in [0, 1]$,

$$f(\tau \overline{x} + (1 - \tau)\overline{y}) \leq \tau f(\overline{x}) + (1 - \tau)f(\overline{y}).$$

Definition 2.2 (See Fig. 2.4.) A set C is *convex* if for all $\overline{x}, \overline{y} \in C$, the convex combination $\tau \overline{x} + (1 - \tau)\overline{y} \in C$ for all $\tau \in (0, 1)$.

Checking whether a given optimization problem is convex is not straightforward in general. Moreover, not all convex problems can be directly solved by the existing software. For these reasons, in practice it is best to focus on well-known forms of convex optimization. For example, linear optimization is convex. Semidefinite

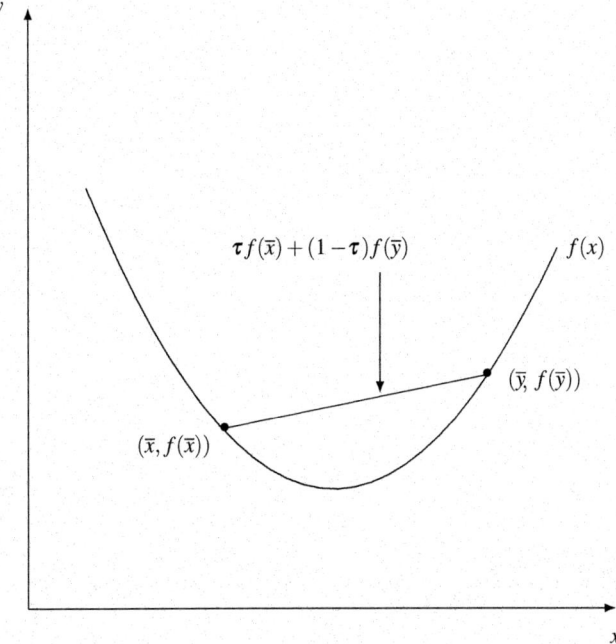

Fig. 2.3 Illustration of the definition of a convex function

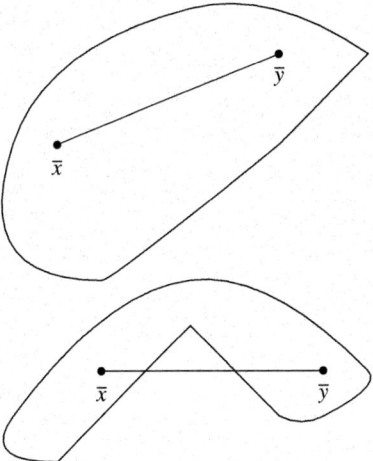

Fig. 2.4 Illustration of the definition of a convex set (above), in contrast to a nonconvex set (below)

optimization and second-order cone optimization are also useful because they are convex and can be solved efficiently using the available software (see Appendix A).

Let us look again at the formulation (2.3–2.5). If the coefficients c_{ij} are non-negative, then the objective function (2.3) is convex because the absolute value is a convex function, and non-negative linear combinations of convex functions are convex. However, the absolute value function is not differentiable everywhere, and handling this type of objective requires specialized nondifferentiable optimization algorithms that are not straightforward to apply.

Furthermore, the constraints (2.4) do not define a convex set. This is also because of the presence of absolute values, and specifically because enforcing a positive lower bound on an absolute value function results in a nonconvex set of solutions. To see why this happens, consider the simple constraint $|x - 1| \geq 1$. The set of values of x satisfying this constraint is $(-\infty, 0] \cap [2, \infty)$, which is not convex.

There are numerous applications of mathematical optimization in which it is difficult or impossible to build a model with a convex objective function and/or a convex feasible set. Various state-of-the-art optimization solvers can be used for nonconvex problems, and some of these solvers could be applied to (2.3–2.5), but they do not provide a guarantee of global optimality.

The best performance is achieved when the objective function and constraints are linear, second-order cone or semidefinite expressions. We demonstrate this for the SRFLP by expressing it using both a mixed-integer linear optimization (MILO) in Sect. 2.3 and rank-constrained semidefinite optimization (SDO) in Sect. 2.7.

2.2 Single-Row Layout as a General Class of Problems

We have illustrated single-row layout using a manufacturing example, but the SRFLP has a broad range of applications. Because the word "machine" is used for a specific type of layout in manufacturing systems, we will use the generic word *department*, as is done for facility layout in general, to refer to the rectangles to be placed in an instance of the SRFLP. Similarly, the space in which the departments are to be arranged is called the *facility*.

Formally, an instance of the SRFLP is defined by a set of n one-dimensional departments $\{1, 2, \ldots, n\}$ with positive lengths $\{\ell_1, \ell_2, \ldots, \ell_n\}$ and non-negative pairwise connectivities c_{ij}. The objective is to find an arrangement of the facilities next to each other in a line so as to minimize the total weighted sum of the centre-to-centre distances between all pairs of facilities, where each distance is weighted by the corresponding connectivity.

The problem can be compactly formulated as

$$\min_{\pi \in \Pi_n} \sum_{i=1}^{n-1} \sum_{j=i+1}^{n} c_{ij} d_{ij}^{\pi}, \qquad (2.6)$$

where Π_n is the set of all permutations π of $\{1, 2, \ldots, n\}$ and d_{ij}^n is the centre-to-centre distance between departments i and j measured with respect to permutation π. This formulation is compact and elegant but not particularly useful in practice. Nevertheless, it provides an important insight, which is that *the SRFLP is an optimization problem over all possible permutations of n departments.*

An important observation here is that if π' denotes the permutation symmetric to π, defined by $\pi'_i = \pi_{n+1-i}$, $i = 1, \ldots, n$, then $d_{ij}^\pi = d_{ij}^{\pi'}$. In other words, for a given layout, *the order of the departments can be reversed without changing the value of the objective function.* This means that it is possible (and important) to apply symmetry-breaking techniques that reduce the computational cost of mathematical optimization algorithms for the SRFLP.

2.3 Mixed-Integer Linear Optimization Approach

Let us recall the formulation (2.3–2.5) but now expressed for n departments:

$$\text{minimize} \quad \sum_{i=1}^{n-1} \sum_{j=i+1}^{n} c_{ij} |x_i - x_j|, \tag{2.7}$$

$$\text{subject to} \quad |x_i - x_j| \geq \frac{1}{2}(\ell_i + \ell_j), \quad i, j = 1, \ldots, n, \ i < j, \tag{2.8}$$

$$\frac{1}{2}\ell_i \leq x_i \leq L - \frac{1}{2}\ell_i, \quad i = 1, \ldots, n. \tag{2.9}$$

In line with the observations in Sect. 2.1.1, we will derive a more efficient formulation that is amenable to practical use.

2.3.1 Linearization of the Distance

We present two approaches to linearizing the objective function (2.7):

- The first approach linearizes $|x_i - x_j|$ by defining new variables $d_{ij} := |x_i - x_j|$ for $i, j = 1, \ldots, n$, $i < j$, rewriting the objective as

$$\min_{x_1,\ldots,x_n} \sum_{i=1}^{n-1} \sum_{j=i+1}^{n} c_{ij} d_{ij}.$$

The objective function is now linear, but we must add constraints on d_{ij}. Simply adding $d_{ij} = |x_i - x_j|$ only shifts the nonlinearity from the objective to the constraints and does not achieve much in practice. We instead take advantage of the non-negativity of the coefficients c_{ij} to linearize the constraints $d_{ij} = |x_i - x_j|$. Observe that because $c_{ij} \geq 0$, we can relax $d_{ij} = |x_i - x_j|$ to $d_{ij} \geq |x_i - x_j|$. This is because no optimal solution has $d_{ij} > |x_i - x_j|$ since it gives an unnecessary increase in the value of the objective function. We can then use the following well-known linearization technique:

$$d_{ij} \geq |x_i - x_j| \quad \Leftrightarrow \quad d_{ij} \geq x_i - x_j \text{ and } d_{ij} \geq x_j - x_i. \tag{2.10}$$

The idea is that because $c_{ij} \geq 0$, by the reasoning above, the optimization will ensure that one of the two linear inequalities on the right-hand side of (2.10) is tight, i.e., holds with equality, at optimality. Unless $x_i = x_j$, only the inequality corresponding to the larger of the two values $x_i - x_j$ and $x_j - x_i$ will be tight, and this larger value is by definition the value of $|x_i - x_j|$.

- The second approach introduces new variables v_{ij}^+ and v_{ij}^- and defines $v_{ij}^+ + v_{ij}^- = |x_i - x_j|$ for $i, j = 1, \ldots, n$, $i < j$. Then the objective function is linearized as

$$\min_{x_1,\ldots,x_n} \sum_{i=1}^{n-1} \sum_{j=i+1}^{n} c_{ij}(v_{ij}^+ + v_{ij}^-),$$

and the following constraints are added to the formulation:

$$v_{ij}^+ \geq 0, \quad v_{ij}^- \geq 0, \text{ and } x_i - x_j + v_{ij}^+ - v_{ij}^- = 0.$$

The idea is that if $x_i - x_j$ is negative, then $v_{ij}^+ = -(x_i - x_j)$ and $v_{ij}^- = 0$. Otherwise, $v_{ij}^+ = 0$ and $v_{ij}^- = (x_i - x_j)$. Since $c_{ij} \geq 0$ and we are minimizing, any other choices of u_{ij}, v_{ij} will increase the objective function and thus cannot be optimal.

2.3.2   Linearization of the Nonoverlap Constraints

We also need to linearize the nonoverlap constraints (2.8), which contain the term $|x_i - x_j|$. Here we cannot exploit the fact that $c_{ij} \geq 0$ in the objective function. Instead we introduce binary variables α_{ij} for $i, j = 1, \ldots, n$, $i < j$ to let the optimization determine the relative position of each pair of departments i and j. In our single-row problem, either i is to the left of j or vice versa. We encode these two possibilities using α_{ij} as follows:

$$\alpha_{ij} = \begin{cases} 1 \text{ if department } i \text{ is to the left of department } j, \\ 0 \text{ if department } j \text{ is to the left of department } i. \end{cases}$$

We now linearize (2.8) as follows:

$$x_i - x_j + M\alpha_{ij} \geq \tfrac{1}{2}(\ell_i + \ell_j) \tag{2.11}$$

$$x_j - x_i + M(1 - \alpha_{ij}) \geq \tfrac{1}{2}(\ell_i + \ell_j), \tag{2.12}$$

where M is a sufficiently large number (see the next subsection for a discussion of how to choose M). These constraints work as follows:

- If $\alpha_{ij} = 0$, then $|x_i - x_j| = x_i - x_j$. In this case, constraint (2.11) becomes

$$x_i - x_j \geq \frac{1}{2}(\ell_i + \ell_j), \text{ and hence } |x_i - x_j| \geq \frac{1}{2}(\ell_i + \ell_j), \text{ as desired.}$$

At the same time, constraint (2.12) has a large value M added to the left-hand side, which means that it will hold regardless of the values of x_i and x_j.
- Similarly, if $\alpha_{ij} = 1$, then $|x_i - x_j| = x_j - x_i$, and constraint (2.12) becomes

$$x_j - x_i \geq \frac{1}{2}(\ell_i + \ell_j), \text{ and hence } |x_i - x_j| \geq \frac{1}{2}(\ell_i + \ell_j), \text{ as desired.}$$

Simultaneously, constraint (2.11) has M added to the left-hand side and in effect imposes no restriction on x_i and x_j.

2.3.3 Choosing M

The so-called big-M method is a modelling technique that uses binary variables to turn otherwise linear constraints on or off. In the previous section, we want to turn one of the nonoverlap constraints on, and the other one off, so that the term $|x_i - x_j|$ is always linearized correctly.

Although the idea is simple, choosing a value for M can be tricky. This is because setting M too small can cause the optimal solution to become infeasible, i.e., it no longer satisfies the constraints. On the other hand, setting M too large can impact the performance of the optimization solver. First, numerical instability can arise when one coefficient (M) is much larger than the others. Second, optimization problems with integer variables are typically solved by combining a branching algorithm (such as branch-and-bound or branch-and-cut) with relaxations where the integer variables are allowed to be continuous. The large value of M almost always makes these relaxations weaker, and weak relaxations can dramatically slow down the solution process.

In general, M should be as small as possible, and whenever possible, knowledge of the application should be used in determining an appropriate value. For constraints (2.11) and (2.12), we can determine the smallest possible (negative) value of the difference between x_i and x_j and choose M accordingly.

Specifically, consider the case $\alpha_{ij} = 1$ with departments i and j placed as far apart as possible in a row of length L, with L as defined earlier. Suppose (without loss of generality) that i is to the left of j. Then $x_i = \frac{1}{2} \ell_i$ and $x_j = L - \frac{1}{2} \ell_j$, and therefore,

$$x_i - x_j = \frac{1}{2}\ell_i - (L - \frac{1}{2} \ell_j) = \frac{1}{2}(\ell_i + \ell_j) - L.$$

We conclude that in this case, $M = L$ is a valid choice for (2.11). By similar reasoning, it follows that this value is also valid for (2.12), and for the case $\alpha_{ij} = 0$.

2.3.4 Mixed-Integer Linear Optimization Formulation

Applying the approaches in Sects. 2.3.1 and 2.3.2 to the formulation (2.7–2.9), we obtain a first MILO model for single-row layout:

$$\text{minimize} \quad \sum_{i<j} c_{ij}(u_{ij} + v_{ij}) \tag{2.13}$$

$$\text{s.t.} \quad x_i - x_j + u_{ij} - v_{ij} = 0, \quad 1 \leq i < j \leq n, \tag{2.14}$$

$$x_i - x_j + L\alpha_{ij} \geq \frac{1}{2}(\ell_i + \ell_j), \quad 1 \leq i < j \leq n, \tag{2.15}$$

$$x_j - x_i + L(1 - \alpha_{ij}) \geq \frac{1}{2}(\ell_i + \ell_j), \quad 1 \leq i < j \leq n, \tag{2.16}$$

$$\frac{1}{2}\ell_i \leq x_i \leq L - \frac{1}{2}\ell_i, \quad 1 \leq i \leq n, \tag{2.17}$$

$$\alpha_{ij} \in \{0, 1\}, \ i = 1, \ldots, n-1, \quad 1 \leq i < j \leq n. \tag{2.18}$$

$$u_{ij}, v_{ij} \geq 0, \quad 1 \leq i < j \leq n. \tag{2.19}$$

This is only one of various MILO formulations of the SRFLP. If we can find a formulation for which the corresponding relaxations are tighter, then branching methods will perform better and also provide tighter global lower bounds on the optimal value.

The next section presents a way to formulate the SRFLP without continuous variables.

2.4 Betweenness-Based Binary Linear Optimization Formulation

The model that we present in this section is based on the following observation: it is not necessary to know the position of each department along the row to formulate the SRFLP. This is because for each pair of departments i and j, it suffices to know which departments are placed *between* i and j. In other words, the key concept for defining a single-row layout is *betweenness*.

This can be observed if we express the SRFLP in the following form:

$$\min_{\pi \in \Pi_n} \sum_{i<j} c_{ij} \left[\frac{1}{2} \left(\ell_i + \ell_j \right) + D_\pi (i, j) \right],$$

where $D_\pi (i, j)$ is the total row space that is taken by the departments located between i and j under permutation π. For example, with the permutation π of 10 departments illustrated in Fig. 2.5, $D_\pi (i, j) = \ell_{k_8} + \ell_{k_1} + \ell_{k_6}$. It follows that $D_\pi (i, j)$ is precisely the sum of the lengths of the departments between i and j under π, under the assumption that the departments are placed with no empty space between them. This assumption will hold if the weights c_{ij} of the objective function are all greater than or equal to zero, as observed earlier, or if the length of the row is $L = \sum_{i=1}^{n} \ell_i$.

To obtain a formulation based on betweenness, instead of defining variables that determine the position of each department or whether a given department is to the left or right of another, we want variables that indicate that a given department is between two others, neither of them necessarily located right next to the given department. We proceed as follows.

For any three distinct departments $i, j, k \in \{1, \ldots, n\}$, define the betweenness variable β_{ijk} as

$$\beta_{ijk} = \begin{cases} 1, & \text{if department } k \text{ lies between departments } i \text{ and } j, \\ 0, & \text{otherwise.} \end{cases}$$

Using these variables, it is straightforward to deduce that $D_\pi (i, j) = \sum_{k \neq i, j} \ell_k \beta_{ijk}$, which is the sum of the lengths of all the departments between i and j. Note that this holds regardless of the relative positions of departments i and j along the row.

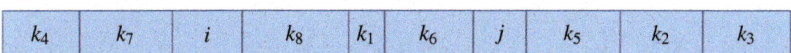

Fig. 2.5 Illustration of betweenness: the distance between i and j equals $\ell_{k_8} + \ell_{k_1} + \ell_{k_6}$

It follows that we can express the objective function of the SRFLP as

$$\sum_{i<j} c_{ij} \left(\frac{1}{2} (\ell_i + \ell_j) + \sum_{k \neq i, j} \ell_k \beta_{ijk} \right).$$

We now need some constraints to ensure that the values assigned to the variables β_{ijk} are consistent. First, for any three distinct departments i, j, k, only one of them is between the other two, and therefore,

$$\beta_{ijk} + \beta_{ikj} + \beta_{jki} = 1, \quad 1 \leq i < j < k \leq n. \tag{2.20}$$

Second, suppose that $\beta_{ikd} = 1$, i.e., department d is between departments i and k. This implies that for every other department j, either d is between i and j or d is between j and k. This observation leads to the constraints:

$$\beta_{ijd} + \beta_{jkd} - \beta_{ikd} \geq 0, \text{ for all } \{i, j, k\} \subseteq \{1, \dots, n\}, \ d \neq i, j, k. \tag{2.21}$$

Third, by similar reasoning, if d is between i and k, then d cannot simultaneously be between i and j and between j and k. This can be expressed as

$$\beta_{ijd} + \beta_{jkd} + \beta_{ikd} \leq 2, \quad 1 \leq i < j < k \leq n, \ d \neq i, j, k. \tag{2.22}$$

Finally, note that the number of betweenness variables can be reduced by half using the observation that

$$\beta_{ijd} = \beta_{jid}, \quad 1 \leq i < j \leq n, \ d \neq i, j.$$

If we apply this reduction in the number of variables, constraints (2.20) and (2.22) are unchanged, but constraint (2.21) must be rewritten as three sets of inequalities:

$$\beta_{ijd} + \beta_{jkd} - \beta_{ikd} \geq 0, \quad 1 \leq i < j < k \leq n, \ d \neq i, j, k, \tag{2.23}$$

$$\beta_{ijd} - \beta_{jkd} + \beta_{ikd} \geq 0, \quad 1 \leq i < j < k \leq n, \ d \neq i, j, k, \tag{2.24}$$

$$-\beta_{ijd} + \beta_{jkd} + \beta_{ikd} \geq 0, \quad 1 \leq i < j < k \leq n, \ d \neq i, j, k. \tag{2.25}$$

In this way, we have assembled all the ingredients for a betweenness-based formulation of the SRFLP:

$$\text{minimize} \quad \sum_{i<j} c_{ij} \left(\frac{1}{2} (\ell_i + \ell_j) + \sum_{k \neq i, j} \ell_k \beta_{ijk} \right) \tag{2.26}$$

$$\text{s.t.} \quad \beta_{ijk} + \beta_{ikj} + \beta_{jki} = 1, \quad 1 \leq i < j < k \leq n, \tag{2.27}$$

$$\beta_{ijd} + \beta_{jkd} - \beta_{ikd} \geq 0, \quad 1 \leq i < j < k \leq n, \ d \neq i, j, k, \tag{2.28}$$

$$\beta_{ijd} - \beta_{jkd} + \beta_{ikd} \geq 0, \quad 1 \leq i < j < k \leq n, \, d \neq i, j, k, \quad (2.29)$$

$$- \beta_{ijd} + \beta_{jkd} + \beta_{ikd} \geq 0, \quad 1 \leq i < j < k \leq n, \, d \neq i, j, k, \quad (2.30)$$

$$\beta_{ijd} + \beta_{jkd} + \beta_{ikd} \leq 2, \quad 1 \leq i < j < k \leq n, \, d \neq i, j, k \quad (2.31)$$

$$\beta_{ijk} \in \{0, 1\}, \quad 1 \leq i < j \leq n, \, k \neq i, j. \quad (2.32)$$

Note that this is a pure binary linear optimization problem.

Most algorithms for binary and integer optimization problems rely on the continuous relaxation, which is the optimization problem obtained when all the integer variables are allowed to vary continuously between their bounds (smallest and largest permitted values). The algorithms use the continuous relaxation to obtain global bounds on the optimal solution, terminating when the best bound matches the objective value of the best solution found so far. If the bounds on the variables can be made tighter, then the continuous relaxation will be tighter as well. More generally, if the feasible set of the continuous relaxation is made smaller (while still containing all the feasible integer solutions), then this usually results in better global bounds, and hence in lower computational times. It turns out that when the binary formulation above is relaxed to a linear optimization problem by allowing the binary variables to be continuous in the interval $0 \leq \beta_{ijk} \leq 1$, the resulting relaxation is weak. A general approach to tighten the continuous relaxation of a given formulation is to add the so-called valid inequalities to the integer optimization problem. The continuous relaxation of the betweenness formulation can be tightened using the class of valid inequalities given below.

Proposition 2.1 *Let $f \leq n$ be a positive even integer and let $S \subseteq \{1, \ldots, n\}$ be such that $|S| = f$. For each $r \in S$, and for any partition (S_1, S_2) of $S \setminus \{r\}$ such that $|S_1| = \frac{1}{2} f$, the inequality*

$$\sum_{\substack{t < q \\ t, q \in S_1}} \beta_{tqr} + \sum_{\substack{t < q \\ t, q \in S_2}} \beta_{tqr} - \sum_{\substack{t \in S_1 \\ q \in S_2}} \beta_{\min\{t, q\}, \max\{t, q\}, r} \leq 0 \quad (2.33)$$

is valid for the betweenness formulation of the SRFLP. □

Proof Let h_1 and h_2 be the number of departments to the left of r in S_1 and S_2, respectively. This means that there are $|S_1| - h_1$ departments in S_1 to the right of r, and similarly $|S_2| - h_2$ departments to the right of r in S_1. Let $(\beta_{ijk})_{1 \leq i < j \leq n, \, i \neq k \neq j}$ be a feasible solution of the SRFLP betweenness formulation. The variable β_{tqr} will be equal to one when:

- t is equal to one of the h_1 departments to the left of r in S_1, and q is equal to one of the $|S_1| - h_1$ other departments in S_1, or
- t is equal to one of the h_2 departments to the left of r in S_2, and q is equal to one of the $|S_2| - h_1$ other departments in S_2.

There are $h_1(|S_1| - h_1)$ such combinations in the first case, and thus, $\sum_{t<q|t,q\in S_1} \beta_{tqr} = h_1(|S_1| - h_1)$. Similarly $\sum_{t<q|t,q\in S_2} \beta_{tqr} = h_2(|S_2| - h_2)$. Hence,

$$\sum_{\substack{t<q \\ t,q\in S_1}} \beta_{tqr} + \sum_{\substack{t<q \\ t,q\in S_2}} \beta_{tqr} = h_1(|S_1| - h_1) + h_2(|S_2| - h_2).$$

On the other hand,

$$\sum_{\substack{t\in S_1 \\ q\in S_2}} \beta_{\min\{t,q\},\max\{t,q\},r} = h_1(|S_2| - h_2) + h_2(|S_1| - h_1).$$

We want to prove that

$$h_1(|S_1| - h_1) + h_2(|S_2| - h_2) - \{h_1(|S_2| - h_2) + h_2(|S_1| - h_1)\} \le 0.$$

Since $|S_2| = |S_1| - 1$, we obtain

$$h_1(|S_1| - h_1) + h_2(|S_1| - 1 - h_2) - \{h_1(|S_1| - 1 - h_2) + h_2(|S_1| - h_1)\} \le 0.$$

This can be simplified to

$$-h_1^2 - h_2 - h_2^2 + h_1 + 2h_1 h_2 \le 0,$$

which is equivalent to

$$h_1 - h_2 \le (h_1 - h_2)^2.$$

Because this is always true for integers h_1 and h_2, the result is proved. □

It is straightforward to check that for $f = 4$, (2.33) is of the form (2.28)–(2.30). To see this, consider (2.33) with $S_1 = \{i, j\}$ and $S_2 = \{k\}$, with $i < j < k < r$. Note that the second sum of (2.33) has no terms because $|S_2| = 1$, and hence (2.33) becomes

$$\beta_{ijr} - \sum_{\substack{t\in S_1 \\ q\in S_2}} \beta_{tqr} \le 0.$$

Therefore, we obtain

$$\beta_{ijr} - \beta_{ikr} - \beta_{jkr} \le 0,$$

which is constraint (2.30).

2.5 Binary Quadratic Optimization Formulation

Betweenness is fundamental to the SRFLP. In the previous section we defined betweenness variables with three indices and modelled the SRFLP using them. We now show that a betweenness-based model can be built using the binary variables α_{ij} defined in Sect. 2.3.2 provided we allow a (nonconvex) quadratic objective function. We also show one way to linearize the resulting formulation and thus obtain a new binary linear optimization model for the SRFLP.

Recall that $\alpha_{ij} = 1$ if department i is to the left of department j, and $\alpha_{ij} = 0$ otherwise. The key observation for a connection to betweenness is that

$\alpha_{ij}\alpha_{jk}$ equals 1 if and only if j is between i and k, and i is to the left of k.

The following theorem shows how to use this observation to express the distance between any two departments i and j as a quadratic function of the variables α_{ij}. We give a short proof that shows the connections with the variables and formulation in Sect. 2.4.

Theorem 2.1 *Let n be the number of departments in an instance of the SRFLP, and ℓ_i denote the length of department i. The centre-to-centre distance between departments i and j in a layout can be expressed using the variables α_{ij} as*

$$\frac{1}{2}(\ell_i + \ell_j) + \sum_{\substack{k=1 \\ k \neq i,j}}^{n} \ell_k \left(\alpha_{ik}\alpha_{kj} + \alpha_{jk}\alpha_{ki}\right), \tag{2.34}$$

with $1 \leq i < j \leq n$.

Proof Note that $\beta_{ijk} = \alpha_{ik}\alpha_{kj} + \alpha_{jk}\alpha_{ki}$, because k is between i and j if and only if k is after i and before k (which makes the term $\alpha_{ik}\alpha_{kj}$ equal to 1), or k is after j or before i (which makes the term $\alpha_{jk}\alpha_{ki}$ equal to 1). Thus,

$$\sum_{k \neq i,j} \ell_k \beta_{ijk} = \sum_{k \neq i,j} \ell_k(\alpha_{ik}\alpha_{kj} + \alpha_{jk}\alpha_{ki}).$$

Substituting this expression into (2.26), we obtain the desired result. □

As we did for the variables β_{ijk} in Sect. 2.4, we need to specify constraints that ensure the consistency of the values assigned to the variables α_{ij}. Note that the constraints that we now derive are not necessary in Sect. 2.3.4 because they are implicitly enforced via the continuous variables x_i. The reasoning we follow is similar, but the constraints are different because the α_{ij} have only two indices, and hence they cannot capture betweenness directly in the way that the variables β_{ijk} do. Nevertheless, the constraints turn out to be simpler than those in Sect. 2.4.

First, we cannot simultaneously have i to the left of j and j to the left of i. Hence, we impose the constraints

$$\alpha_{ij} + \alpha_{ji} = 1, \quad 1 \leq i < j \leq n.$$

Second, for any three distinct departments i, j, k, we enforce the transitivity condition:

if i is to the left of j and j is to the left of k, then i is to the left of k.

In other words, if $\alpha_{ij} = 1$ and $\alpha_{jk} = 1$, then we must have $\alpha_{ik} = 1$ (or $\alpha_{ki} = 0$, by 2.36). Equivalently, not all of α_{ij}, α_{jk}, and α_{ki} may be equal to 1 simultaneously. We express this using the constraints

$$\alpha_{ij} + \alpha_{jk} + \alpha_{ki} \leq 2, \quad 1 \leq i \neq j \neq k \leq n.$$

We thus obtain a binary quadratic formulation of the SRFLP:

$$\text{minimize} \quad \sum_{i<j} c_{ij} \left(\frac{1}{2}(\ell_i + \ell_j) + \sum_{\substack{k=1 \\ k \neq i,j}}^{n} \ell_k(\alpha_{ik}\alpha_{kj} + \alpha_{jk}\alpha_{ki}) \right) \tag{2.35}$$

$$\text{s.t.} \quad \alpha_{ij} + \alpha_{ji} = 1, \quad 1 \leq i < j \leq n, \tag{2.36}$$

$$\alpha_{ij} + \alpha_{jk} + \alpha_{ki} \leq 2, \quad 1 \leq i \neq j \neq k \leq n, \tag{2.37}$$

$$\alpha_{ij} \in \{0, 1\}, \quad 1 \leq i \neq j \leq n. \tag{2.38}$$

We note that the number of α_{ij} variables can be reduced by substituting half of them using constraints (2.36). While state-of-the-art optimization solvers may do this automatically, it is best to perform this step before handing over the model to the solver.

Formulation (2.35)–(2.38) is challenging to solve in practice, primarily because of the nonconvex quadratic objective function. We now outline ways to linearize it, leading to either a binary linear optimization problem (Sect. 2.6) or a binary semidefinite optimization problem (Sect. 2.7).

2.6 Linearizing the Binary Quadratic Optimization Formulation

2.6.1 Initial Linearization

We can linearize the formulation (2.35)–(2.38) by introducing new variables representing pairwise products of the α_{ij} variables. Consider any three distinct departments i, j, k. For every such choice of three departments, we define new continuous variables

$$Y_{pq} = \alpha_p \alpha_q, \text{ for } p, q \in \{ij, ji, ik, ki, jk, kj\}. \tag{2.39}$$

With these new variables, it is straightforward to linearize the objective function (2.35):

$$\sum_{i<j} c_{ij} \left(\frac{1}{2}(\ell_i + \ell_j) + \sum_{\substack{k=1 \\ k \neq i,j}}^{n} \ell_k (Y_{ikkj} + Y_{jkki}) \right). \tag{2.40}$$

Next we must link the continuous variables Y with the binary variables α using only linear constraints. This can be achieved with the following set of constraints:

$$Y_{ijjk} \geq -1 + \alpha_{ij} + \alpha_{jk}, \quad Y_{ijjk} \leq \alpha_{ij}, \quad Y_{ijjk} \leq \alpha_{jk}, \quad 1 \leq i,j,k \leq n, \tag{2.41}$$

$$Y_{ijik} \geq -1 + \alpha_{ij} + \alpha_{ik}, \quad Y_{ijik} \leq \alpha_{ij}, \quad Y_{ijik} \leq \alpha_{ik}, \quad 1 \leq i,j,k \leq n, \tag{2.42}$$

$$Y_{ikjk} \geq -1 + \alpha_{jk} + \alpha_{ik}, \quad Y_{ikjk} \leq \alpha_{ik}, \quad Y_{ikjk} \leq \alpha_{jk}, \quad 1 \leq i,j,k \leq n. \tag{2.43}$$

Each set of constraints works as follows: If both α variables are equal to 1, then the corresponding Y variable will also be equal to 1 because of the first inequality. On the other hand, if at least one of the α variables is equal to 0, then the Y variable will also equal 0 because of the second and third inequalities.

Let us work through the two cases of (2.41) in detail. First, suppose that i is to the left of j and j is to the left of k, so that $\alpha_{ij} = 1$ and $\alpha_{jk} = 1$. Then $Y_{ijjk} \geq -1 + \alpha_{ij} + \alpha_{jk} = -1 + 1 + 1 = 1$, and hence $Y_{ijjk} \geq 1$. At the same time, $Y_{ijjk} \leq \alpha_{ij} = 1$, and hence $Y_{ijjk} \leq 1$. Thus, $Y_{ijjk} = 1$ must hold. Alternatively, suppose that i is to the left of j and k is to the left of j, so that $\alpha_{ij} = 1$ and $\alpha_{jk} = 0$. Then $Y_{ijjk} \geq -1 + \alpha_{ij} + \alpha_{jk} = -1 + 1 + 0 = 0$, and hence $Y_{ijjk} \geq 0$. At the same time, $Y_{ijjk} \leq \alpha_{jk} = 0$, and hence $Y_{ijjk} \leq 0$. Thus, $Y_{ijjk} = 0$ must hold.

An important additional observation about constraints (2.41)–(2.43) is that if all the α variables are binary, then the variables Y will all take on binary values, even if we define them as continuous variables. Finally, observe that $Y_{ijkd} = Y_{kdij}$ by definition, so we add the constraints $Y_{ijkd} = Y_{kdij}$ for $1 \leq i \neq j, k \neq d \leq n$.

This leads us to the following linearization of the quadratic integer program (2.35)–(2.38):

$$\text{minimize} \quad \sum_{i<j} c_{ij} \left(\frac{1}{2}(\ell_i + \ell_j) + \sum_{\substack{k=1 \\ k \neq i,j}}^{n} \ell_k \left(Y_{ikkj} + Y_{jkki} \right) \right) \tag{2.44}$$

$$\text{s.t.} \quad Y_{ijjk} \geq -1 + \alpha_{ij} + \alpha_{jk}, \quad Y_{ijjk} \leq \alpha_{ij}, \quad Y_{ijjk} \leq \alpha_{jk}, \quad 1 \leq i,j,k \leq n, \tag{2.45}$$

$$Y_{ijik} \geq -1 + \alpha_{ij} + \alpha_{ik}, \quad Y_{ijik} \leq \alpha_{ij}, \quad Y_{ijik} \leq \alpha_{ik}, \quad 1 \leq i, j, k \leq n,$$
$$\tag{2.46}$$

$$Y_{ikjk} \geq -1 + \alpha_{jk} + \alpha_{ik}, \quad Y_{ikjk} \leq \alpha_{ik}, \quad Y_{ikjk} \leq \alpha_{jk}, \quad 1 \leq i, j, k \leq n,$$
$$\tag{2.47}$$

$$\alpha_{ij} + \alpha_{ji} = 1, \quad 1 \leq i < j \leq n, \tag{2.48}$$

$$\alpha_{ij} + \alpha_{jk} + \alpha_{ki} \leq 2, \quad 1 \leq i \neq j \neq k \leq n, \tag{2.49}$$

$$Y_{ijkd} = Y_{kdij}, \quad 1 \leq i \neq j, k \neq d \leq n, \tag{2.50}$$

$$\alpha_{ij} \in \{0, 1\}, \quad 1 \leq i \neq j \leq n. \tag{2.51}$$

As we observed for the binary quadratic formulation, the number of α variables can be reduced by substituting half of them using (2.48). The same is true for the Y variables, based on constraint (2.50). This reduction is useful for computational efficiency, and it should be carried out before the model is handed over to the solver.

2.6.2   Improving the Linearized Formulation

As already noted, valid inequalities are essential for the efficient solution of mathematical optimization models. The linearized formulation (2.44)–(2.51) can be improved by the addition of valid inequalities.

Valid inequalities for the SRFLP and other problems can often be generated using a framework known as the *reformulation–linearization technique (RLT)*. The simple but extremely powerful process behind RLT can be summarized in two steps:

1. The reformulation step introduces new terms involving the product of two (or more) variables by multiplying one or more of the constraints by one or more of the variables.
2. The linearization step replaces each new product term by a new variable and links the new variables to the former variables using linear constraints.

We illustrate here a straightforward application of RLT but emphasize that the framework is highly flexible and can be used in different ways.

Before applying RLT, we revisit constraints (2.49) and observe that for each choice of distinct i, j, k, two different 3-cycles are forbidden, namely $i \rightarrow j \rightarrow k \rightarrow i$ and $i \rightarrow k \rightarrow j \rightarrow i$. Therefore, we can split each of the constraints into two inequalities:

$$\alpha_{ij} + \alpha_{jk} + \alpha_{ki} \leq 2 \text{ and } \alpha_{ik} + \alpha_{kj} + \alpha_{ji} \leq 2, \quad 1 \leq i \neq j \neq k \leq n.$$

Without loss of generality, if we assume that $i < j < k$, then from (2.48) we have $\alpha_{ki} = 1 - \alpha_{ik}$, $\alpha_{kj} = 1 - \alpha_{jk}$, and $\alpha_{ji} = 1 - \alpha_{ij}$. Substituting using these gives

$$\alpha_{ij} + \alpha_{jk} - \alpha_{ik} \leq 1 \text{ and } \alpha_{ik} - \alpha_{jk} - \alpha_{ij} \leq 0, \quad 1 \leq i < j < k \leq n.$$

For clarity of presentation, let us write these two inequalities as a single (vector) inequality:

$$\begin{bmatrix} 1 \\ -1 \end{bmatrix} \alpha_{ij} + \begin{bmatrix} 1 \\ -1 \end{bmatrix} \alpha_{jk} + \begin{bmatrix} -1 \\ 1 \end{bmatrix} \alpha_{ik} \leq \begin{bmatrix} 1 \\ 0 \end{bmatrix}. \tag{2.52}$$

We now follow the RLT steps. First, we multiply both sides of (2.52) by α_{ij} and obtain

$$\begin{bmatrix} 1 \\ -1 \end{bmatrix} \alpha_{ij}^2 + \begin{bmatrix} 1 \\ -1 \end{bmatrix} \alpha_{jk}\alpha_{ij} + \begin{bmatrix} -1 \\ 1 \end{bmatrix} \alpha_{ik}\alpha_{ij} \leq \begin{bmatrix} 1 \\ 0 \end{bmatrix} \alpha_{ij}.$$

Second, we linearize the result using the variables Y as well as the fact that $\alpha_{ij}^2 = \alpha_{ij}$ and obtain

$$\begin{bmatrix} 1 \\ -1 \end{bmatrix} Y_{ijjk} + \begin{bmatrix} -1 \\ 1 \end{bmatrix} Y_{ijik} \leq \begin{bmatrix} 0 \\ 1 \end{bmatrix} \alpha_{ij}. \tag{2.53}$$

In general, new variables may need to be defined at this step, but we have simply used the Y variables defined previously.

We can also multiply both sides of (2.52) by $(1 - \alpha_{ij})$ and linearize, to obtain

$$\begin{bmatrix} -1 \\ 1 \end{bmatrix} Y_{ijjk} + \begin{bmatrix} 1 \\ -1 \end{bmatrix} Y_{ijik} + \begin{bmatrix} 1 \\ -1 \end{bmatrix} \alpha_{jk} + \begin{bmatrix} -1 \\ 1 \end{bmatrix} \alpha_{ik} + \begin{bmatrix} 1 \\ 0 \end{bmatrix} \alpha_{ij} \leq \begin{bmatrix} 1 \\ 0 \end{bmatrix}. \tag{2.54}$$

The inequalities (2.53) and (2.54) for all $1 \leq i < j < k \leq n$ can be used to tighten (2.44)–(2.51). Furthermore, many other inequalities can be generated using RLT. The difficulty in practice is that adding valid inequalities also significantly increases the computational cost of solving the relaxation. It is therefore important to strike a balance between the number of valid inequalities added and the resulting computational cost. We discuss this in more detail in Sect. 2.8.

2.7 Semidefinite Optimization Formulation

In this section we look at an alternative way to linearize the binary quadratic formulation (2.35)–(2.38), namely using semidefinite optimization instead of linear optimization. We provide a short introduction to SDO in Appendix A.

We start by defining a set of binary variables γ_{ij} that will perform exactly the same role as the α_{ij} variables defined in Sect. 2.3.2, but with the two possible values being $\{-1, 1\}$ instead of $\{0, 1\}$. While the whole SDO approach can be applied using $\{0, 1\}$, we make this change because in SDO it is much more convenient to use the values $\{-1, 1\}$ than the usual choice of $\{0, 1\}$ in LO. We define γ_{ij} as follows:

$$\gamma_{ij} = \begin{cases} 1 & \text{if department } i \text{ is to the left of department } j, \\ -1 & \text{if department } i \text{ is to the right of department } j. \end{cases}$$

We immediately observe that $\gamma_{ij} = -\gamma_{ji}$, so we can easily avoid duplication of variables by requiring that $i < j$, and changing the sign as well as the order of the indices whenever $i > j$ occurs.

It is important to point out that in the SDO context, we can use both linear and quadratic constraints, and indeed it is best to use quadratic expressions as much as possible to take full advantage of the power of SDO. This means that our approach to modelling problems using SDO is often quite different to the LO approach. In this spirit, we choose to express the binary nature of the γ variables using the quadratic constraints

$$\gamma_{ij}^2 = 1 \text{ for all } i < j.$$

Let us now develop the constraints necessary to ensure the consistency of the values assigned to γ_{ij}, following the reasoning applied in Sect. 2.5 for α_{ij}.

First, we note that $\gamma_{ij} = -\gamma_{ji}$ corresponds precisely to the relationship $\alpha_{ij} = 1 - \alpha_{ji}$ in the $\{0, 1\}$ formulations, and hence for the γ variables it is automatically true that we cannot simultaneously have i to the left of j and j to the left of i.

Second, for any three distinct departments i, j, k, we want to ensure that the transitivity condition holds:

if i is to the left of j and j is to the left of k, then i is to the left of k.

Observe that this condition is equivalent to

$$\text{if } \gamma_{ij} = \gamma_{jk} \text{ then } \gamma_{ik} = \gamma_{ij}. \tag{2.55}$$

It can be formulated using the following quadratic constraint:

$$(\gamma_{ij} + \gamma_{jk})(\gamma_{ij} - \gamma_{ik}) = 0. \tag{2.56}$$

The reasoning here is that this constraint requires one of the two expressions in parentheses to be equal to zero. If the first expression is zero, then $\gamma_{ij} \neq \gamma_{jk}$, and there is nothing more to do. If the first expression is not zero, then $\gamma_{ij} = \gamma_{jk}$ and the second expression must be zero, which means that $\gamma_{ik} = \gamma_{ij}$ must hold, precisely as desired. Expanding the expression in (2.56), we can write the constraint in the form

$$\gamma_{ij}\gamma_{jk} - \gamma_{ij}\gamma_{ik} - \gamma_{ik}\gamma_{jk} = -1, \ 1 \leq i < j < k \leq n, \tag{2.57}$$

after also noting that all the permutations of i, j, k lead to this same quadratic expression once expanded.

We next turn to the objective function (2.35) and seek to express betweenness using the γ_{ij} variables. This is straightforward after we observe that department k is between i and j if and only if $\gamma_{ki}\gamma_{kj} = -1$, and if and only if $\frac{1-\gamma_{ki}\gamma_{kj}}{2} = 1$. Using this observation, we can express (2.35) in terms of the γ variables:

$$\sum_{i<j} c_{ij} \left(\frac{1}{2}(\ell_i + \ell_j) + \sum_{\substack{k=1 \\ k\neq i,j}}^{n} \ell_k \left(\frac{1-\gamma_{ki}\gamma_{kj}}{2} \right) \right).$$

If we want to avoid the occurrence of indices ij with $i > j$, using our earlier observation that $\gamma_{ij} = -\gamma_{ji}$ allows us to rewrite the objective function in the following form:

$$\left(\sum_{i<j} \frac{c_{ij}}{2} \right) \left(\sum_{k=1}^{n} \ell_k \right) - \sum_{i<j} \frac{c_{ij}}{2} \left[\sum_{k<i} \ell_k \gamma_{ki}\gamma_{kj} - \sum_{i<k<j} \ell_k \gamma_{ik}\gamma_{kj} + \sum_{k>j} \ell_k \gamma_{ik}\gamma_{jk} \right].$$

We can now write a fully quadratic formulation of the SRFLP:

$$\text{minimize} \quad \left(\sum_{i<j} \frac{c_{ij}}{2} \right) \left(\sum_{k=1}^{n} \ell_k \right) - \sum_{i<j} \frac{c_{ij}}{2} \left[\sum_{k<i} \ell_k \gamma_{ki}\gamma_{kj} - \sum_{i<k<j} \ell_k \gamma_{ik}\gamma_{kj} + \sum_{k>j} \ell_k \gamma_{ik}\gamma_{jk} \right]$$

$$\tag{2.58}$$

$$\text{s.t.} \quad \gamma_{ij}\gamma_{jk} - \gamma_{ij}\gamma_{ik} - \gamma_{ik}\gamma_{jk} = -1, \, 1 \leq i < j < k \leq n, \tag{2.59}$$

$$\gamma_{ij}^2 = 1, \, 1 \leq i < j \leq n. \tag{2.60}$$

We next use this fully quadratic formulation to obtain a rank-constrained SDO formulation of the SRFLP. Since SDO is an optimization problem over matrices, we need to define a matrix variable. Let us define the column vector

$$g := (\gamma_{12}, \gamma_{13}, \ldots, \gamma_{1n}, \gamma_{23}, \ldots, \gamma_{n-1\,n})^T,$$

and hence the rank-one matrix

$$\Gamma := g\, g^T. \tag{2.61}$$

By construction, we have that $\Gamma_{p,q} = \gamma_p\gamma_q$ for every pair p, q of departments. Note that the matrix Γ is indexed by pairs of departments, like the variable Y in Sect. 2.6.1. Using this fact, we can linearize the formulation (2.58)–(2.60) by expressing every quadratic term in terms of entries of Γ.

It remains to determine how to handle the requirement that the matrix Γ be of the form (2.61). We make the following observations:

- First, observe that Γ is by construction a rank-one matrix. This is because every column of Γ is a multiple of the vector g; in fact, each column is equal to either g or $-g$. We denote this by rank $(\Gamma) = 1$.
- Second, observe that because each γ_{ij} is equal to -1 or 1, all the entries of Γ are also equal to -1 or 1. It follows that the diagonal entries of Γ are equal to 1 (because they are of the form γ_{ij}^2). We denote this by diag $(\Gamma) = e$, where e denotes the vector of all ones (of the appropriate dimension in the context).
- Third, by Theorem A.2 in Appendix A, we have that Γ is positive semidefinite (PSD). We denote this by $\Gamma \succeq 0$.

It turns out that these three conditions suffice to ensure that Γ has the form (2.61). Therefore, we can formulate the SRFLP as follows:

$$\text{minimize} \quad \left(\sum_{i<j} \frac{c_{ij}}{2}\right)\left(\sum_{k=1}^{n} \ell_k\right) - \sum_{i<j} \frac{c_{ij}}{2}\left[\sum_{k<i} \ell_k \Gamma_{ki,kj} - \sum_{i<k<j} \ell_k \Gamma_{ik,kj} + \sum_{k>j} \ell_k \Gamma_{ik,jk}\right]$$

$$\tag{2.62}$$

$$\text{s.t.} \quad \Gamma_{ij,jk} - \Gamma_{ij,ik} - \Gamma_{ik,jk} = -1, \, 1 \le i < j < k \le n, \tag{2.63}$$

$$\text{rank } (\Gamma) = 1, \tag{2.64}$$

$$\text{diag } (\Gamma) = e, \tag{2.65}$$

$$\Gamma \succeq 0. \tag{2.66}$$

This formulation is nearly an SDO problem. The only constraint that cannot be handled by SDO solvers is (2.64) because it restricts the rank of the matrix variables. This constraint can be interpreted as an "integrality" constraint, in the sense that among all the matrices that are PSD and have all diagonal entries equal to 1, those whose entries are all integer (equal to -1 or 1), are precisely those that have rank equal to 1. It is therefore natural to *relax* this constraint (analogously to relaxations in MILO) by replacing it with bounds on the entries of Γ:

$$-1 \le \Gamma_{p,q} \le 1.$$

Conveniently, it turns out that these bounds are implied by the combination of the constraints diag $(\Gamma) = e$ and $\Gamma \succeq 0$. Indeed, by (A.2) in Appendix A, we have that $\Gamma_{p,q}^2 \le 1$ for all pairs p, q. Hence, relaxing the rank constraint in the formulation above comes down to simply removing it.

The resulting SDO relaxation of the SRFLP is

$$\text{minimize} \quad \left(\sum_{i<j} \frac{c_{ij}}{2}\right)\left(\sum_{k=1}^{n} \ell_k\right) - \sum_{i<j} \frac{c_{ij}}{2}\left[\sum_{k<i} \ell_k \Gamma_{ki,kj} - \sum_{i<k<j} \ell_k \Gamma_{ik,kj} + \sum_{k>j} \ell_k \Gamma_{ik,jk}\right]$$

$$\tag{2.67}$$

$$\text{s.t.} \quad \Gamma_{ij,jk} - \Gamma_{ij,ik} - \Gamma_{ik,jk} = -1, \, 1 \le i < j < k \le n, \tag{2.68}$$

$$\text{diag}\,(\Gamma) = e, \tag{2.69}$$

$$\Gamma \succeq 0. \tag{2.70}$$

Note that unless the optimal solution matrix Γ^* has rank equal to 1, we obtain only a lower bound on the optimal value of the SRFLP, and not a feasible solution. Therefore, as in MILO, we are interested in ways to tighten the relaxation, and in general we may need to apply an enumeration algorithm to obtain the global optimal solution.

We observed earlier that we can reverse the order of the departments in the SRFLP without changing the value of the objective function. This reversal of the order is equivalent to replacing every γ_{ij} value by its negative. We observe that if we do this, then there is no change to the SDO formulation or its relaxation, because all the expressions involved are quadratic in the γ variables. In this way, the SDO formulation and the corresponding SDO relaxation implicitly account for the symmetry of the SRFLP.

2.7.1   Improving the Semidefinite Formulation

The SDO formulation can be improved by the addition of valid inequalities, as was done for the LO formulation. In particular, the RLT framework can be applied to generate valid inequalities for the entries of Γ. Because the SDO relaxation is in practice already very tight, it turns out that a single class of valid inequalities is highly effective. These inequalities are known as the *triangle inequalities*, and they are well known in combinatorial optimization.

For our purposes, we focus on the interpretation that the triangle inequalities model the fact that for any assignment of the values $\{-1, 1\}$ to the variables γ, the values of the matrix entries $\Gamma_{p,q}$, $\Gamma_{p,r}$, and $\Gamma_{q,r}$, where p, q, r are any three distinct pairs of departments, must comprise an even number of negative values. In other words, either none of these three matrix entries equal -1, or precisely two of them equal -1. (It is straightforward to verify this claim by checking all possible cases.) This demonstrates the validity of the triangle inequalities. For $\{-1, 1\}$ binary variables, these inequalities take the following form:

$$\Gamma_{p,q} + \Gamma_{p,r} + \Gamma_{q,r} \geq -1 \tag{2.71}$$

$$\Gamma_{p,q} - \Gamma_{p,r} - \Gamma_{q,r} \geq -1 \tag{2.72}$$

$$-\Gamma_{p,q} - \Gamma_{p,r} + \Gamma_{q,r} \geq -1 \tag{2.73}$$

$$-\Gamma_{p,q} + \Gamma_{p,r} - \Gamma_{q,r} \geq -1, \tag{2.74}$$

where p, q, r are any three distinct pairs of departments.

For the specific case of the SDO relaxation of SRFLP, we observe that not all of these inequalities will help to tighten the relaxation. Specifically, suppose that the pairs p, q, r are equal to $ij, ik,$ and jk for any three departments i, j, k. Then it is straightforward to check that the inequalities (2.71)–(2.74) are implied by constraint (2.68). Therefore, the $4\binom{n}{3}$ triangle inequalities arising from the choices of pairs of the form $\{p, q, r\} = \{ij, ik, jk\}$ cannot improve the SDO relaxation. However, for choices of pairs p, q, r that involve four or more departments, the triangle inequalities can help significantly.

Another class of valid inequalities that has been useful for tightening the SDO relaxation of SRFLP is

$$-1 - \gamma_{st} \leq \gamma_{ij} + \gamma_{jk} - \gamma_{ik} + \Gamma_{ij,st} + \Gamma_{jk,st} - \Gamma_{ik,st} \leq 1 + \gamma_{st}, \tag{2.75}$$

$$-1 + \gamma_{st} \leq \gamma_{ij} + \gamma_{jk} - \gamma_{ik} - \Gamma_{ij,st} - \Gamma_{jk,st} + \Gamma_{ik,st} \leq 1 - \gamma_{st}, \tag{2.76}$$

for all $1 \leq i < j < k \leq n, 1 \leq s < t \leq n$. These inequalities were obtained using the Lovász–Schrijver (LS) procedure, which is analogous to the RLT framework. A fundamental idea in these frameworks is to exploit the fact that any binary solution must satisfy the nonlinear equation $\gamma_{ij}^2 = \gamma_{ij}$ for all $1 \leq i < j \leq n$. We do not give the details here and refer the reader to Sect. 2.10 for further information.

2.8 Inequality Separation

We have presented several classes of inequalities applicable to the SRFLP. We have also mentioned two procedures to generate such classes for the SRFLP, and indeed for general optimization problems, namely RLT and LS. The reader may suspect (or already know) that the number of valid inequalities available is very large. In general it is not possible or desirable to add every known valid inequality to a relaxation, whether it is an LO or an SDO problem.

Consider the triangle inequalities (2.71)–(2.74). There are $4\binom{\binom{n}{2}}{3}$ of them, which means their number is $O(n^6)$. We can omit the $4\binom{n}{3}$ implied inequalities, but we still have far too many to add them all to the SDO relaxation. Moreover, given a specific instance of the SRFLP, the vast majority of them will not be active at the optimal solution and hence do not help to tighten the relaxation. This is true for most classes of inequalities of interest. It is therefore essential to have a way to identify which inequalities within each class will be useful for a given relaxation. This process is called *separation*.

For a given class of inequalities, a *separation algorithm* takes a given point P as input and gives one of two possible outputs:

- one inequality in the given class that is not satisfied by P, or
- the conclusion that no such inequality exists, i.e., the point P satisfies all the inequalities in the class.

Note that not all separation algorithms are efficient; we are particularly interested in those that run in polynomial time. Effective separation algorithms are an essential component of the most successful approaches for solving instances of the SRFLP and other facility layout problems. For small values of n, the classes of inequalities that we have presented can often be separated by full enumeration because the inequalities are all given explicitly. Given a current solution, it is straightforward to check whether each inequality is violated, and if so, to compute the amount of the violation. Some of the most violated inequalities can then be added to the relaxation. For larger values of n, either heuristic or more sophisticated separation algorithms must be used.

2.9 SRFLP with Departments of Equal Length

We now consider the special case of the SRFLP in which all the departments have the same length. This is also called the equidistant SRFLP or the single-row equidistant facility layout problem (SREFLP) because it involves placing the departments in a given set of equally spaced locations along a straight line. We can assume without loss of generality that $\ell_i = 1$ for all departments i, and all the models in this chapter can be directly applied to the SREFLP.

The minimum-backtracking row layout problem (MBRLP) is a version of the SREFLP that arises in the sequencing of machines along an automated production line. The quantity of workflow from one machine to another is a directed quantity, and the ideal sequencing is one in which there is no need for materials to flow backwards along the production line. This is often not possible, and we wish to find the sequencing of machines that minimizes the total distance backtracked. This turns out to be an SREFLP with a different objective function.

The formal statement of the MBRLP is as follows. Given n departments and directed flows f_{ij} from i to j for $i, j = 1, \ldots, n$, where f_{ij} and f_{ji} are different in general and $f_{ii} = 0$, the objective is to find a one-to-one assignment of the departments to n locations equally spaced along a straight line so as to minimize the total weighted backward distance, where each pairwise backward distance is weighted by its corresponding flow.

We again use the variables α from earlier sections: $\alpha_{ij} = 1$ if department i is located to the left of department j, and $\alpha_{ij} = 0$ otherwise. We fix the desired direction of flow from left to right, so that backward flow is any flow from right to left. In other words, the flow from i to j is *backwards* if j is to the left of i, i.e., if $\alpha_{ji} = 1$.

Let d_{ij}^b denote the backtracking distance from machine i to machine j, $i \neq j$, $i, j = 1, \ldots, n$. Note that we always have one of d_{ij}^b and d_{ji}^b equal to zero and the other positive, depending on the relative positions of i and j. Observe also that the backtracking distance between i and j is exactly the number of positions between them. We use this observation to express the backtracking distance. First, suppose that department j is to the left of department i. Then the number of positions between them can be computed by counting the number of departments to the left of i and subtracting the number of departments to the left of j. We thus have

$$d_{ij}^b = \sum_{k \neq i} \alpha_{ki} - \sum_{k \neq j} \alpha_{kj}.$$

Similarly, if i is to the left of j, then the backtracking distance is

$$d_{ji}^b = \sum_{k \neq j} \alpha_{kj} - \sum_{k \neq i} \alpha_{ki}.$$

We can express both cases simultaneously as follows:

$$d_{ij}^b - d_{ji}^b = \sum_{k \neq i} \alpha_{ki} - \sum_{k \neq j} \alpha_{kj}$$

$$d_{ij}^b, d_{ji}^b \geq 0.$$

The idea is that depending on the sign of the difference on the right-hand side, one of d_{ij}^b and d_{ji}^b will be positive and the other zero. Naturally, they could both be positive and have the correct difference between them to equate to the right-hand side, but this will not happen because of the objective function, which is clearly

$$\min_{d_{ij}^b} \sum_{i=1}^{n} \sum_{j \neq i} f_{ij} d_{ij}^b.$$

This minimization will make all the d_{ij}^b variables as small as possible, so the desired result is obtained.

The formulation of the MBRLP is thus

$$\text{minimize} \quad \sum_{i=1}^{n} \sum_{j \neq i} f_{ij} d_{ij}^b \tag{2.77}$$

$$\text{s.t.} \quad d_{ji}^b - d_{ij}^b = \sum_{k \neq i} \alpha_{ki} - \sum_{k \neq j} \alpha_{kj}, \quad 1 \leq i < j \leq n, \tag{2.78}$$

$$d_{ij}^b \geq 0, \quad 1 \leq i \neq j \leq n, \tag{2.79}$$

$$\alpha_{ij} + \alpha_{ji} = 1, \quad 1 \leq i < j \leq n, \tag{2.80}$$

$$\alpha_{ij} + \alpha_{jk} + \alpha_{ki} \leq 2, \quad 1 \leq i \neq j \neq k \leq n, \tag{2.81}$$

$$\alpha_{ij} \in \{0, 1\}, \quad 1 \leq i \neq j \leq n, \tag{2.82}$$

where we use the constraints on the α variables derived in Sect. 2.5.

2.10   References and Further Reading

The example of Tables 2.1 and 2.2 was adapted from Heragu (2008). Boyd and Vandenberghe (2004) is an excellent reference on the theory, algorithms, and applications of convex optimization. The compact formulation (2.6) was expressed in Simmons (1969), where the centrality of betweenness was observed for the first time.

The first MILO formulations of the SRFLP were proposed by Love and Wong (1976) and Heragu and Kusiak (1991). Amaral (2006) proposed a MILO formulation that gives a tighter linear relaxation than the Heragu and Kusiak (1991) formulation. This is presented in Sect. 2.3.4.

The approach sketched in Sect. 2.4, where the betweenness variables are used, was originally proposed in Amaral (2009). A polyhedral study of this formulation can be found in Sanjeevi and Kianfar (2010).

The quadratic integer model was introduced in Amaral (2008). The quadratic integer model of Sect. 2.5 has an objective function inspired by the integer linear model of Sect. 2.4: it is the model of Amaral (2008) with a different objective function. The relaxation hierarchy discussed here was introduced in Sherali and Adams (1990, 1994), and Adams and Sherali (2015). It can be obtained by applying the relaxation described in Adams and Sherali (2015) to the quadratic integer model of Sect. 2.5.

Section 2.7 presents the first SDO approach for the SRFLP as introduced in Anjos et al (2005). The combination of the SDO approach and triangle inequalities was used in Anjos and Vannelli (2008) to compute global optimal layouts for single-row facility layout problems with up to 30 facilities. An interesting study of the SDO approach for one-dimensional layout problems is given by Hungerländer and Rendl (2013). Sufficient conditions for a matrix to be of the form (2.61) and related properties of PSD matrices with all diagonal entries equal to 1 are discussed in, for example, Wolkowicz and Anjos (2002). The SDO approach is to date the best approach for solving large instances of the SRFLP.

Valid inequalities such as (2.75) can be obtained using the Lovász–Schrijver procedure introduced in Lovász and Schrijver (1991). An excellent presentation of this procedure as well as other frameworks for constructing hierarchies of linear and semidefinite relaxations for binary problems can be found in Laurent and Rendl (2005).

The SREFLP can be viewed as a special case of the famous quadratic assignment problem. Some work has focused on exploiting its special structure, most recently Palubeckis (2012) and Hungerländer (2014). A formulation of the MBRLF was given in Kouvelis and Chiang (1996), but the computation of the backward distance was incorrect, as explained in Brusco (2004), where a corrected formulation is provided. The formulation presented in Sect. 2.9 follows the approach of Brusco (2004).

References

Adams WP, Sherali HD (2015) RLT insights into lift-and-project closures. Optimization Letters 9(1):19–39

Amaral ARS (2006) On the exact solution of a facility layout problem. Eur J Oper Res 173(2):508–518

Amaral ARS (2008) An exact approach to the one-dimensional facility layout problem. Operations Research 56(4):1026–1033

Amaral ARS (2009) A new lower bound for the single row facility layout problem. Discrete Appl Math 157(1):183–190

Anjos MF, Vannelli A (2008) Computing globally optimal solutions for single-row layout problems using semidefinite programming and cutting planes. INFORMS J Comput 20(4):611–617

Anjos MF, Kennings A, Vannelli A (2005) A semidefinite optimization approach for the single-row layout problem with unequal dimensions. Discrete Optimization 2(2):113–122

Boyd SP, Vandenberghe L (2004) Convex optimization. Cambridge University Press

Brusco MJ (2004) Optimal solution methods for the minimum-backtracking row layout problem. IIE Transactions 36(2):181–189

Heragu SS (2008) Facilities design. CRC Press

Heragu SS, Kusiak A (1991) Efficient models for the facility layout problem. Eur J Oper Res 53(1):1–13

Hungerländer P (2014) Single-row equidistant facility layout as a special case of single-row facility layout. Int J Prod Res 52(5):1257–1268

Hungerländer P, Rendl F (2013) A computational study and survey of methods for the single-row facility layout problem. Comput Optim Appl 55(1):1–20

Kouvelis P, Chiang WC (1996) Optimal and heuristic procedures for row layout problems in automated manufacturing systems. J Oper Res Soc 47(6):803–816

Laurent M, Rendl F (2005) Semidefinite programming and integer programming. Handbooks Oper Res Manag Sci 12:393–514

Lovász L, Schrijver A (1991) Cones of matrices and set-functions and 0–1 optimization. SIAM J Optim 1(2):166–190

Love R, Wong J (1976) On solving a one-dimensional space allocation problem with integer programming. INFOR Inf Syst Oper Res 14(2):139–143

Palubeckis G (2012) A branch-and-bound algorithm for the single-row equidistant facility layout problem. OR Spectrum 34(1):1–21

Sanjeevi S, Kianfar K (2010) A polyhedral study of triplet formulation for single row facility layout problem. Discrete Appl Math 158(16):1861–1867

Sherali HD, Adams WP (1990) Relaxations between the continuous and convex hull representations. SIAM J Discrete Math 3(3):411–430

Sherali HD, Adams WP (1994) A hierarchy of relaxations and convex hull characterizations for mixed-integer zero-one programming problems. Discrete Appl Math 52(1):83–106

Simmons DM (1969) One-dimensional space-allocation algorithm: An ordering algorithm. Operations Research 17(5):812–826

Wolkowicz H, Anjos MF (2002) Semidefinite programming for discrete optimization and matrix completion problems. Discrete Appl Math 123(1–2):513–577

Chapter 3
Layout on Several Rows

In this chapter we consider the problem of optimally placing a set of departments on two or more rows, given the lengths of the departments, the number of rows, and a pairwise non-negative connectivity for each pair of departments. We refer to such problems as row layout problems, and we divide this class of problems into two types, namely double-row facility layout and multi-row facility layout, where the latter involves three or more rows. It is worth treating double-row facility layout separately because it is amenable to specialized models that are typically more efficient than those for multi-row layout.

Facility layout problems on rows arise in manufacturing systems, and the type of layout depends on the choice of material handling system. While single-row and double-row layouts use the same type of system, an AGV, layouts on three or more rows are of interest when using a gantry robot.

Another application is in the design of some integrated circuits for which the layout of the components is organized in rows (called base layers). The objective is to minimize the total wirelength required to connect the components, where the space between the rows is used for the wires connecting the components.

A separation between the rows is often specified to account for factors such as the movements of the material handling system and/or the workers. This is usually straightforward to accommodate in an optimization model. A more challenging aspect is that within each row, a minimum clearance between departments is normally required to meet safety and operational requirements. For simplicity, we assume that this clearance is included in the lengths of the departments, but it can also be handled by adjusting the nonoverlap constraints to enforce the clearance.

In this chapter we assume that the rows and departments all have the same height, that every department can be assigned to every row, and that the distances between

© Springer Nature Switzerland AG 2021
M. F. Anjos, M. V. C. Vieira, *Facility Layout*, EURO Advanced Tutorials
on Operational Research, https://doi.org/10.1007/978-3-030-70990-7_3

adjacent rows are equal. Under these assumptions, solving an instance of facility layout on two or more rows involves three related tasks:

1. assign each department to exactly one of the rows;
2. express mathematically the weighted centre-to-centre distance between pairs of departments (which may or may not be in the same row); and
3. account for the possibility of empty space between departments in the same row.

Clearly, both double-row and multi-row layouts are more challenging than single-row layout because in the latter there is no need to assign departments to rows (there is only one row), and $c_{ij} \geq 0$ ensures that there is no empty space between departments at optimality, so that empty space generally does not need to be modelled. Hence, the focus in the models of Chap. 2 was for the most part on different ways to mathematically express the centre-to-centre distance between departments within a row. By contrast, for the models in this chapter we must carry out all three tasks.

3.1 Double-Row Facility Layout

The double-row facility layout problem (DRFLP) requires the departments to be placed on both sides of a central corridor. The idea here is that the flows between departments can be handled by an AGV that travels back-and-forth along the corridor, similarly to the case of the SRFLP. In this context, the distance between the two rows is considered negligible, and thus the centre-to-centre distance between two departments is measured along the corridor. Figure 3.1 illustrates the DRFLP with the corridor as the operating space for an AGV.

Despite this similarity with the SRFLP, the DRFLP is much more challenging to model and to solve. On the one hand, if we know which departments are placed in one of the rows, it follows that the remaining departments are in the other row. On the other hand, betweenness information is no longer sufficient to determine centre-to-centre distances, and furthermore the optimal layout may have some empty space between departments.

In this section we describe two approaches that extend in different ways the MILO formulations presented in Chap. 2 for the SRFLP. Both extensions involve a combination of discrete and continuous variables, where the discrete variables represent the assignment of departments to rows and the relative position of pairs of departments, and the continuous variables give the positions of the department

Fig. 3.1 DRFLP with a corridor for an AGV

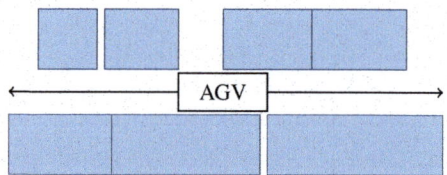

centres with respect to a fixed origin. Without loss of generality, we place the corridor along the x-axis, with the origin at the left end of the corridor.

3.1.1 Initial Mixed-Integer Linear Optimization Model

We consider again the binary variables α_{ij}, $1 \leq i, j \leq n$, used in Chap. 2, but we change their definition. Specifically, we extend them to represent not only the relative position of departments i and j but also whether or not these two departments are in the same row. The definition of the α_{ij} is therefore as follows:

$$\alpha_{ij} = \begin{cases} 1, \text{ if department } i \text{ is to the left of department } j \\ \quad \text{ and both } i \text{ and } j \text{ are in the same row;} \\ 0, \text{ otherwise.} \end{cases}$$

Note that $\alpha_{ij} = 0$ whenever i and j are not in the same row and also when they are in the same row but i is to the right of j.

We naturally need constraints to ensure that the values assigned to the variables α_{ij} are consistent. The first two sets of constraints discussed below are analogous to the models for single row, and the third and fourth cover situations that cannot occur in the SRFLP. First, we cannot simultaneously have i to the left of j and j to the left of i (whether or not they are in the same row). Hence, we have the constraints

$$\alpha_{ij} + \alpha_{ji} \leq 1, \quad 1 \leq i < j \leq n. \tag{3.1}$$

Second, we need transitivity constraints with respect to row assignments. Specifically, we need to ensure that if departments i and k are in the same row and k and j are in the same row, then i and j are also in the same row. We model this transitivity requirement via the constraints

$$\alpha_{ik} + \alpha_{ki} + \alpha_{jk} + \alpha_{kj} - \alpha_{ij} - \alpha_{ji} \leq 1, \ 1 \leq i, j, k \leq n, \ i < j, \ k \neq i, j. \tag{3.2}$$

These constraints work as follows. If i and j are in the same row, then $\alpha_{ik} + \alpha_{ki} = 1$. Similarly, if k and j are in the same row, then $\alpha_{jk} + \alpha_{kj} = 1$. Whenever both of these hold, constraint (3.2) takes the form

$$1 \geq \alpha_{ik} + \alpha_{ki} + \alpha_{jk} + \alpha_{kj} - \alpha_{ij} - \alpha_{ji} = 1 + 1 - (\alpha_{ij} + \alpha_{ji}),$$

which implies that $\alpha_{ij} + \alpha_{ji} \geq 1$ must hold, i.e., i and j must be in the same row.

Third, we need constraints to prevent the (impossible) situation where k is to the left of j, j is to the left of i, and i is to the left of k. These so-called three-cycle constraints are as follows:

$$\alpha_{ik} + \alpha_{ji} + \alpha_{kj} - \alpha_{ki} - \alpha_{ij} - \alpha_{jk} \leq 1, \ 1 \leq i, j, k \leq n, \ i, k < j, \ k \neq i. \tag{3.3}$$

Note that if all three terms are equal to 1, then this inequality requires (at least) one of the other terms to be equal to 1 as well, but this contradicts (3.1).

Fourth, we require that for every triple i, j, k of departments, at least two must be in the same row. We model this via the constraints

$$\alpha_{ij} + \alpha_{ik} + \alpha_{jk} + \alpha_{ji} + \alpha_{ki} + \alpha_{kj} \geq 1, \quad \{i, j, k\} \subset \{1, \ldots, n\}. \tag{3.4}$$

These constraints ensure that no more than two rows are used, as required by the DRFLP.

We also introduce the continuous variables x_i, $1 \leq i \leq n$, and d_{ij}, $1 \leq i < j \leq n$, with x_i representing the position of the centre of department i along the corridor and d_{ij} representing the distance between the centres of i and j measured along the corridor. These quantities are the same as those in Chap. 2 for the SRFLP.

Using the variables and constraints described above, we can write an initial mixed-integer linear formulation of the DRFLP as follows:

$$\text{minimize} \quad \sum_{i=1}^{n-1} \sum_{j=i+1}^{n} c_{ij} d_{ij} \tag{3.5}$$

$$\text{s.t.} \quad d_{ij} \geq x_i - x_j, \; d_{ij} \geq x_j - x_i, \quad 1 \leq i < j \leq n, \tag{3.6}$$

$$x_i + \left(\frac{\ell_i + \ell_j}{2} \right) \leq x_j + L(1 - \alpha_{ij}), \quad 1 \leq i \neq j \leq n, \tag{3.7}$$

$$d_{ij} - \left(\frac{\ell_i + \ell_j}{2} \right) \alpha_{ij} - \left(\frac{\ell_i + \ell_j}{2} \right) \alpha_{ji} \geq 0, \quad 1 \leq i < j \leq n, \tag{3.8}$$

(3.1), (3.2), (3.3), (3.4),

$$\alpha_{ij} \in \{0, 1\}, \quad 1 \leq i \neq j \leq n, \tag{3.9}$$

$$\frac{\ell_i}{2} \leq x_i \leq L - \frac{\ell_i}{2}, \quad 1 \leq i \leq n, \tag{3.10}$$

where $L = \sum_{i=1}^{n} \ell_i$. In this formulation, constraints (3.6) give the centre-to-centre distance between each pair of departments using the same approach as for the SRFLP. Also similarly to the SRFLP, constraints (3.7) ensure that departments assigned to the same row do not overlap. Constraints (3.8) ensure that if department i is placed in the same row as department j, then the distance between their centres is at least $(\ell_i + \ell_j)/2$. Constraints (3.9) require the α_{ij} variables to be binary, and constraints (3.10) are the same bounds on the x variables as those used for the SRFLP.

We observe that constraints (3.1) and (3.8) are redundant in the presence of constraints (3.6) and (3.7), but they may be helpful in improving the performance of a branching algorithm.

3.1.2   Improved Mixed-Integer Linear Optimization Model with Betweenness Variables

We can improve formulation (3.6)–(3.10) by making use of betweenness information. Specifically, we can change the right-hand side of constraints (3.8) from 0 to a positive number based on betweenness and thus improve the lower bound obtained when solving the LO relaxations via a branching algorithm. To achieve this, we extend the definition of betweenness variables, originally given in Sect. 2.4. For three distinct departments i, j, and k, define the betweenness variables β_{ijk} as

$$\beta_{ijk} = \begin{cases} 1, & \text{if department } k \text{ lies between departments } i \text{ and } j \\ & \text{and } i, j, \text{ and } k \text{ are all in the same row;} \\ 0, & \text{otherwise.} \end{cases}$$

Starting from the formulation derived in Sect. 3.1.1 and using these new betweenness variables, we obtain the following improved formulation of the DRFLP:

$$\text{minimize} \quad \sum_{i=1}^{n-1} \sum_{j=i+1}^{n} c_{ij} d_{ij} \tag{3.11}$$

$$\text{s.t.} \quad d_{ij} \geq x_i - x_j, \ d_{ij} \geq x_j - x_i, \quad 1 \leq i < j \leq n, \tag{3.12}$$

$$x_i + \left(\frac{\ell_i + \ell_j}{2}\right) \leq x_j + L(1 - \alpha_{ij}), \quad 1 \leq i \neq j \leq n, \tag{3.13}$$

$$d_{ij} - \left(\frac{\ell_i + \ell_j}{2}\right)\alpha_{ij} - \left(\frac{\ell_i + \ell_j}{2}\right)\alpha_{ji} \geq \sum_{k \neq i,j} \ell_k \beta_{kij}, \quad 1 \leq i < j \leq n \tag{3.14}$$

$$(3.1) - (3.4), \tag{3.15}$$

$$\beta_{ijk} \geq \alpha_{ik} + \alpha_{kj} - 1, \quad 1 \leq i < j \leq n, \ k \neq i, j, \tag{3.16}$$

$$\beta_{ijk} \geq \alpha_{ki} + \alpha_{jk} - 1, \quad 1 \leq i < j \leq n, \ k \neq i, j, \tag{3.17}$$

$$\beta_{ijk} \leq \alpha_{ij} + \alpha_{ji}, \quad 1 \leq i < j \leq n, \ k \neq i, j, \tag{3.18}$$

$$\beta_{ijk} \leq \alpha_{ik} + \alpha_{ki}, \quad 1 \leq i < j \leq n, \ k \neq i, j, \tag{3.19}$$

$$\beta_{ijk} \leq \alpha_{jk} + \alpha_{kj}, \quad 1 \leq i < j \leq n, \ k \neq i, j, \tag{3.20}$$

$$\alpha_{ij}, \beta_{ijk} \in \{0, 1\}, \quad 1 \leq i \neq j \leq n, \ k \neq i, j, \tag{3.21}$$

$$\frac{\ell_i}{2} \leq x_i \leq L - \frac{\ell_i}{2}, \quad 1 \leq i \leq n. \tag{3.22}$$

The new constraints (3.16)–(3.20) determine the values of the variables β_{ijk} for each feasible assignment to the variables $\alpha_{ij}, \alpha_{ik}, \alpha_{jk}, \alpha_{ji}, \alpha_{ki}$, and α_{kj}. Specifically, whenever k is between i and j, then constraints (3.16) and (3.17) enforce $\beta_{ijk} = 1$. Alternatively, if one of $\{i, j, k\}$ is in a different row, constraints (3.18)–(3.20) ensure that $\beta_{ijk} = 0$.

3.2 Multi-Row Facility Layout

An instance of the multi-row facility layout problem (MRFLP) has a given number of rows to which the departments can be assigned. The departments all have the same height (equal to the row height), the distances between adjacent rows are equal, and departments can in general be assigned to any row. While from a mathematical point of view, the MRFLP is a natural extension of single-row or double-row layout, from an industrial engineering perspective, layout on three or more rows is a more challenging problem.

The matter of measuring the rectilinear distance between the centres of two departments takes on a greater importance when considering layouts on three or more rows. When two departments are assigned to different rows separated by at least one other row, clearly a component of this distance arises from the separation between the rows. This is not the case for one- and two-row versions of layout, where only the horizontal component of the distance matters.

The MRFLP captures the basic structure of applications where the departments are to be arranged in well-defined rows because the separation between the rows is prespecified. It is thus a problem that is discrete in one dimension (between rows) and continuous in the other dimension (within rows).

We first introduce a straightforward nonlinear model as an extension of the single-row layout model (2.3)–(2.5):

$$\text{minimize} \quad \sum_{i=1}^{n-1} \sum_{j=i+1}^{n} c_{ij}(|x_i - x_j| + |y_i - y_j|) \tag{3.23}$$

$$\text{s.t.} \quad |x_i - x_j| \geq \frac{1}{2}(\ell_i + \ell_j) - L(1 - \alpha_{ij}), \quad 1 \leq i < j \leq n, \tag{3.24}$$

$$|y_i - y_j| \geq \frac{1}{2}(w_i + w_j) - L\alpha_{ij}, \quad 1 \leq i < j \leq n, \tag{3.25}$$

$$\alpha_{ij} \in \{0, 1\}, \quad 1 \leq i < j \leq n, \tag{3.26}$$

where w_i is the width of department i, and for each pair $i < j$,

$$\alpha_{ij} = \begin{cases} 1, & \text{if departments } i \text{ and } j \text{ are in the same row} \\ 0, & \text{otherwise.} \end{cases}$$

This model allows departments to have different widths, but the width of each row must be (at least) the largest width of the departments it contains. If we assume that every department can be assigned to every row, then we may simplify the problem by considering all the departments to have the same width, and this is straightforward to formulate using MILO.

3.2.1 Initial Mixed-Integer Linear Optimization Model

For this model, we define two sets of binary variables:

$$y_{ik} = \begin{cases} 1, & \text{if department } i \text{ is assigned to row } k \\ 0, & \text{otherwise.} \end{cases}$$

$$z_{kij} = \begin{cases} 1, & \text{if department } j \text{ is placed to the right of department } i \text{ in row } k \\ 0, & \text{otherwise.} \end{cases}$$

As in the previous model, we use continuous variables to determine the location of the departments. Specifically, we let x_{ik} denote the (absolute) location of department i in row k, setting it to zero if i is not assigned to row k. We use up to m rows to place the departments. The formulation is as follows:

$$\text{minimize} \quad \sum_{i=1}^{n-1} \sum_{j=i+1}^{n} c_{ij} \left(v_{ij}^+ + v_{ij}^- \right)$$

$$\text{s.t.} \quad \sum_{k=1}^{m} x_{ik} - \sum_{k=1}^{m} x_{jk} + v_{ij}^+ - v_{ij}^- = 0, \quad 1 \le i < j \le n \tag{3.27}$$

$$x_{ik} \le L y_{ik}, \quad i = 1, \ldots, n, \ k = 1, \ldots, m, \tag{3.28}$$

$$\sum_{k=1}^{m} y_{ik} = 1, \quad i = 1, \ldots, n \tag{3.29}$$

$$\frac{\ell_i y_{ik} + \ell_j y_{jk}}{2} \le x_{ik} - x_{jk} + L(1 - z_{kji}), \quad 1 \le i < j \le n, \ k = 1, \ldots, m \tag{3.30}$$

$$\frac{\ell_i y_{ik} + \ell_j y_{jk}}{2} \le x_{jk} - x_{ik} + L(1 - z_{kij}), \quad 1 \le i < j \le n, \ k = 1, \ldots, m \tag{3.31}$$

$$z_{kij} + z_{kji} \leq y_{ik}, \quad 1 \leq i < j \leq n, \ k = 1, \ldots, m, \tag{3.32}$$

$$z_{kij} + z_{kji} \leq y_{jk}, \quad 1 \leq i < j \leq n, \ k = 1, \ldots, m, \tag{3.33}$$

$$z_{kij} + z_{kji} + 1 \geq y_{ik} + y_{jk}, \quad 1 \leq i < j \leq n, \ k = 1, \ldots, m, \tag{3.34}$$

$$\begin{aligned}
&x_{ik} \geq 0, \quad i = 1, \ldots, n, \ k = 1, \ldots, m, \\
&v_{ij}^+, v_{ij}^- \geq 0, \quad 1 \leq i < j \leq n, \\
&y_{ik} \in \{0, 1\}, \quad i = 1, \ldots, n, \ k = 1, \ldots, m, \\
&z_{kij} \in \{0, 1\}, \quad 1 \leq i \neq j \leq n, k = 1, \ldots, m.
\end{aligned} \tag{3.35}$$

Constraints (3.27) compute the distances between departments using the second linearization approach described in Sect. 2.3.1. Constraints (3.28) set $x_{ik} = 0$ when department i is not assigned to row k. Constraints (3.29) ensure that a department is assigned to just one row. Constraints (3.30) and (3.31) prevent departments from overlapping if they are located in the same row. Constraints (3.32)–(3.34) ensure consistency between the variables y and z as follows: if $y_{ik} = 1$ and $y_{jk} = 1$, then (3.32)–(3.34) ensure that exactly one of z_{kij} and z_{kji} is equal to one. Otherwise, at least one of y_{ik} or y_{jk} is equal to zero, and one of (3.32) and (3.33) sets both z_{kij} and z_{kji} to zero.

We point out that this model is designed for an arbitrary number of rows m, and therefore it can be applied to double-row layout ($m = 2$). However, specialized models for that case are normally more efficient than this general model.

3.2.2   Mixed-Integer Linear Optimization Model with Continuous Row Assignments

The formulation in this section is a natural extension of the model (3.23)–(3.26): the variable y_i that indicates the row assignment for each department is now continuous. Like the model presented in the previous subsection and most other mathematical optimization formulations, this model uses binary variables to prevent overlap. Unlike most other models, however, it uses continuous variables for the assignment of departments to rows because it can be proved that these variables will always attain integer values at optimality; the proof is presented in Sect. 3.2.3. This means that the departments are assigned to rows without the need for rounding or a similar operation.

For each department i, we use the variable x_i to represent its horizontal position (within the row it is assigned to) and the variable y_i to represent its vertical position (the row it is assigned to). For each pair of departments i and j, we use the following binary variables to encode their relative positions:

$$\alpha_{ij} = \begin{cases} 1 \text{ if } i \text{ is placed to the left of } j \text{ in the same row} \\ 0 \text{ otherwise,} \end{cases}$$

$$\beta_{ij} = \begin{cases} 1 \text{ if } i \text{ and } j \text{ are placed in different rows and } i \text{ is below } j \\ 0 \text{ otherwise.} \end{cases}$$

Let d_{ij}^x and d_{ij}^y represent the horizontal and vertical distances between i and j. Using the above variable definitions, the formulation for the MRFLP is

$$\text{minimize} \quad \sum_{1 \le i < j \le n} c_{ij}(d_{ij}^x + d_{ij}^y) \tag{3.36}$$

$$\text{s.t.} \quad d_{ij}^x \ge x_i - x_j, \ d_{ij}^x \ge x_j - x_i, \quad 1 \le i < j \le n \tag{3.37}$$

$$d_{ij}^y \ge y_i - y_j, \ d_{ij}^y \ge y_j - y_i, \quad 1 \le i < j \le n \tag{3.38}$$

$$x_j - x_i \ge \frac{1}{2}(\ell_i + \ell_j) - L(1 - \alpha_{ij}), \quad 1 \le i < j \le n, \tag{3.39}$$

$$x_i - x_j \ge \frac{1}{2}(\ell_i + \ell_j) - L(1 - \alpha_{ji}), \quad 1 \le i < j \le n, \tag{3.40}$$

$$y_j - y_i \ge d^{\text{row}} - md^{\text{row}}(1 - \beta_{ij}), \quad 1 \le i < j \le n, \tag{3.41}$$

$$y_i - y_j \ge d^{\text{row}} - md^{\text{row}}(1 - \beta_{ji}), \quad 1 \le i < j \le n, \tag{3.42}$$

$$y_i - y_j \le (1 - \alpha_{ij} - \alpha_{ji})(m - 1)d^{\text{row}}, \quad 1 \le i < j \le n, \tag{3.43}$$

$$y_j - y_i \le (1 - \alpha_{ij} - \alpha_{ji})(m - 1)d^{\text{row}}, \quad 1 \le i < j \le n, \tag{3.44}$$

$$0 \le y_i \le (m - 1)d^{\text{row}}, \quad 1 \le i \le n, \tag{3.45}$$

$$\frac{1}{2}\ell_i \le x_i \le L - \frac{1}{2}\ell_i, \quad 1 \le i \le n, \tag{3.46}$$

$$\alpha_{ij} + \alpha_{ji} + \beta_{ij} + \beta_{ji} = 1, \quad 1 \le i < j \le n, \tag{3.47}$$

$$\alpha_{ij} + \alpha_{jk} \le 1 + \alpha_{ik}, \quad 1 \le i, j, k \le n, \ i < j, \ k \ne i, j \tag{3.48}$$

$$\beta_{ij} + \beta_{jk} \le 1 + \beta_{ik}, \quad 1 \le i, j, k \le n, \ i < j, \ k \ne i, j \tag{3.49}$$

$$\alpha_{ij}, \beta_{ij} \in \{0, 1\}, \quad 1 \le i, j \le n, \tag{3.50}$$

where d^{row} is the row width (common to all rows), and as before, n is the number of departments, m is the maximum number of rows allowed for the layout, ℓ_i is the length of department i, and $L = \sum_{i=1}^{n} \ell_i$.

Constraints (3.37)–(3.38) establish the horizontal and vertical distances between departments using the first linearization approach described in Sect. 2.3.1. Constraints (3.39)–(3.40) prevent any two departments in the same row from overlapping. Constraints (3.41)–(3.42) avoid the overlapping of rows and simultaneously create the rows. Constraints (3.43)–(3.44) ensure that $y_i = y_j$ when departments i and j are placed in the same row. Constraints (3.45) provide bounds on the variables y_i, and in particular they restrict every feasible solution to have no more than m rows (each of width d). Constraints (3.46) provide bounds on the variables x_i. Constraints (3.47) require the separation of i and j in one of the two dimensions (they may be separated in both dimensions at optimality). Constraints (3.48) and (3.49) are transitivity constraints, as seen earlier. Finally, constraints (3.50) require the α and β variables to be binary.

We observe that constraints (3.45) do not guarantee an optimal solution with exactly m rows but rather at most m rows. Thus, this model for the MRFLP is more general than that in Sect. 3.2.1 because it decides the optimal number of rows; most other models require the number of rows to be prespecified.

3.2.3 Proof of the Integrality of Row Assignments at Optimality

As mentioned above, the formulation presented in Sect. 3.2.2 uses continuous variables y_i to represent the vertical position of each department, i.e., the assignment of departments to rows. It turns out that at optimality, the variables y_i always take on integer values, and hence the row assignments are well defined. The proof of this fact uses the concept of *total unimodularity*. We first state the required theoretical concepts and then give the proof.

Definition 3.1 A matrix A with integer elements is *totally unimodular (TU)* if the determinant of each square submatrix of A is equal to 0, 1, or -1.

Example 3.1 The matrix

$$\begin{bmatrix} 0 & 1 & 1 & 0 \\ -1 & 0 & 0 & 0 \\ 0 & 0 & -1 & 1 \end{bmatrix}$$

is TU, but the matrix

$$\begin{bmatrix} 0 & 1 & 1 & 0 \\ -1 & 0 & 0 & 0 \\ 0 & 1 & -1 & 1 \end{bmatrix}$$

is not (because the determinant of the submatrix $\begin{bmatrix} 1 & 1 \\ 1 & -1 \end{bmatrix}$ equals -2).

We say that a polyhedron is integer if all its extreme points have integer coordinates. If we optimize a linear function over an integer polyhedron, then there exists an optimal integer solution (unless the optimization is unbounded). TU is a very useful property because of the following fact.

Proposition 3.1 *If A is TU and b is a vector of integers, then the set of solutions of the system $Ax \leq b$ is either empty or an integer polyhedron.*

The more difficult part is deciding whether or not a given matrix is TU. To apply the TU definition, it is necessary to compute the determinants of all the square submatrices, and their number is exponential in the dimensions of the matrix. For this reason, we are interested in criteria for TU that can be checked efficiently (i.e., in polynomial time). It is straightforward to deduce from Definition 3.1 that every element of a TU matrix must be equal to 0, 1, or -1 and that if A is TU, then so is its transpose A^T. Another useful criterion is given by the following proposition.

Proposition 3.2 *If a given matrix A has all its elements equal to 0, 1, or -1, if it has no more than two nonzero elements in each column, and if for each column with two nonzeros the sum of the elements of that column equals zero, then A is TU.*

Note that Proposition 3.2 makes it easy to check that the first matrix of Example 3.1 is TU. We can now state and prove the property that the optimal values of the variables y_i in (3.36)–(3.50) are always integer.

Theorem 3.1 *For every instantiation of the variables α_{ij} and β_{ij} that satisfies (3.47)–(3.50), the coefficient matrix of the constraints (3.39)–(3.46) is TU.*

Proof Let A be the coefficient matrix of the constraints (3.39)–(3.46). Observe that every element in A^T is $-1, 0$, or 1, i.e., the coefficients of the variables y_i are $-1, 0$, or 1. Since every column of A^T corresponds to the coefficients of y_i, y_j in each constraint, it contains at most two nonzero elements. Where a column of A^T has two nonzero elements, these are of opposite sign. Therefore, by Proposition 3.2, A^T is TU. Thus, A is also TU.   □

Corollary 3.1 *For every feasible instantiation of the variables α_{ij} and β_{ij}, if d is integer, then the y-components of every extreme point of (3.39)–(3.46) are integer.*

Proof For integer d, the right-hand side of the constraints (3.41)–(3.45) is integer. The result follows by Theorem 3.1 and Proposition 3.1.   □

3.2.4 Alternative Optimization Approaches for the MRFLP

The MRFLP is indeed a challenging problem. As a consequence, beyond the MILO approaches in Sects. 3.2.1 and 3.2.2 that directly model the entire problem, the MRFLP has successfully been modelled using alternative approaches that consider different aspects of the problem in turn.

Recall the three tasks (stated at the beginning of this chapter) that arise for an instance of MRFLP:

1. assign each department to exactly one of the rows;
2. express mathematically the weighted centre-to-centre distance between pairs of departments (which may or may not be in the same row);
3. account for the possibility of empty space between departments in the same row.

The idea is to have simpler models that do not perform all these tasks but can be used as part of an overall strategy for the MRFLP.

A successful example of such an approach is as follows. Suppose that the assignment of departments to rows (the first task above) is known, and set up a model that optimizes the layout for that given assignment. Then by enumerating all possible assignments of departments to rows and optimizing the layout for each of the assignments, one can solve the MRFLP. Naturally, if the enumeration is done by brute force, then there will be no gain in efficiency compared to the MILO models above. However, if we enumerate the assignments in a clever way, then we obtain a more efficient way to solve the MRFLP.

The remaining sections of this chapter consider special cases of the MRFLP, some of which have been used in such alternative approaches.

3.3 Fixed-Row Multi-Row Facility Layout

In this section we consider the special case of fixed-row MRFLP, abbreviated FR-MRFLP. The FR-MRFLP is the variant of MRFLP in which the assignment of departments to rows is fixed and given as input. Thus, the optimization does not need to assign each department to a row.

To model the FR-MRFLP, we assume that we are given as input the set $R = \{1, 2, \ldots, m\}$ of rows for the layout and a description of the row assignments in the form of sets $N_r, r \in R$, of the departments assigned to row r.

For the MILO model below, we introduce two dummy departments $n + 1$ and $n + 2$ to be placed at the left and right ends, respectively, of the layout. These dummy departments have zero lengths, i.e., $\ell_{n+1} = \ell_{n+2} = 0$, and they do not contribute any cost to the objective function, so we set $c_{ij} = 0$ if $i \in \{n + 1, n + 2\}$ or $j \in \{n + 1, n + 2\}$. Each department set N_r is extended to include the dummy departments, with the notation $\tilde{N}_r = N_r \cup \{n + 1, n + 2\}$.

The MILO model presented, unlike those in Sects. 3.2.1 and 3.2.2, does not use continuous variables for the position of the departments within rows. Instead it is built on the betweenness model from Sect. 2.4. Therefore, we recall the betweenness variables defined in Sect. 2.4 for any three distinct departments i, j, and k:

$$\beta_{ijk} = \begin{cases} 1, & \text{if department } k \text{ lies between departments } i \text{ and } j, \\ 0, & \text{otherwise.} \end{cases}$$

The MILO model is

$$\text{minimize} \quad \sum_{\substack{i,j \in N \\ i < j}} c_{ij} d_{ij} \tag{3.51}$$

$$\text{s.t.} \quad \beta_{ijk} + \beta_{ikj} + \beta_{jki} = 1, \quad r \in R, \ i,j,k \in \tilde{N}_r, \ i < j < k, \tag{3.52}$$

$$\beta_{ijh} + \beta_{ikh} - \beta_{jkh} \geq 0, \quad r \in R, \ i,j,k,h \in \tilde{N}_r, \ h \neq i,j,k, \ i < j < k, \tag{3.53}$$

$$\beta_{ijh} - \beta_{ikh} + \beta_{jkh} \geq 0, \quad r \in R, \ i,j,k,h \in \tilde{N}_r, \ h \neq i,j,k, \ i < j < k, \tag{3.54}$$

$$-\beta_{ijh} + \beta_{ikh} + \beta_{jkh} \geq 0, \quad r \in R, \ i,j,k,h \in \tilde{N}_r, \ h \neq i,j,k, \ i < j < k, \tag{3.55}$$

$$\beta_{ijh} + \beta_{ikh} + \beta_{jkh} \leq 2, \quad r \in R, \ i,j,k,h \in \tilde{N}_r, \ h \neq i,j,k, \ i < j < k, \tag{3.56}$$

$$\beta_{n+1,n+2,i} = 1, \quad i \in N, \tag{3.57}$$

$$\beta_{ij,n+1} = 0, \quad r \in R, \ i,j, \in N_r \cup \{n+2\}, \ i < j, \tag{3.58}$$

$$\beta_{ij,n+2} = 0, \quad r \in R, \ i,j, \in N_r \cup \{n+1\}, \ i < j, \tag{3.59}$$

$$\beta_{j,n+1,i} = \beta_{i,n+2,j}, \quad r \in R, \ i,j \in N_r, \ i \neq j, \tag{3.60}$$

$$d_{j,n+1} - d_{i,n+1} \geq \frac{\ell_i + \ell_j}{2} + L\,(\beta_{j,n+1,i} - 1), \quad r \in R, \ i,j \in N_r, \ i \neq j, \tag{3.61}$$

$$d_{ij} \geq d_{i,n+1} - d_{j,n+1}, \quad i,j \in N, \ i < j, \tag{3.62}$$

$$d_{ij} \geq d_{j,n+1} - d_{i,n+1}, \quad i,j \in N, \ i < j, \tag{3.63}$$

$$d_{i,n+1} \geq \ell_i / 2, \quad i \in N, \tag{3.64}$$

$$\beta_{ijk} \in \{0,1\}, \quad r \in R, \ i,j,k \in N_r \cup \{n+1, n+2\}, \ i < j, \tag{3.65}$$

where $L = \sum_{i=1}^{n} \ell_i$.

Constraints (3.52)–(3.56) are exactly the same as constraints (2.27)–(2.31) in Sect. 2.4, and we refer the reader to that section for an explanation of their meaning.

Constraints (3.57)–(3.60) ensure that in each row, all the departments lie between the dummy departments $n+1$ and $n+2$. Constraints (3.57) say that every department must be placed between the dummy departments $n+1$ and $n+2$. Constraints (3.58) and (3.59) prevent the dummy departments from being placed between any other two departments. Constraints (3.60) say that i is between j and $n+1$ if and only if j is between i and $n+2$.

Constraints (3.61) give the distance between each pair of departments in the same row: if i is between j and $n+1$, then $\beta_{j,n+1,i} = 1$, which implies $d_{j,n+1} - d_{i,n+1} \geq \ell_i + \ell_j$. These constraints prevent any two departments, placed in the same row, from overlapping.

Constraints (3.62) and (3.63) define the distances between all pairs of depart-ments. Note that these constraints will be effective only for pairs of departments that are not in the same row (because if two departments are in the same row, their distance will be set by constraints (3.61)).

Constraints (3.64) give a trivial lower bound for the distance between the departments and the left boundary of the layout, and constraints (3.65) require the betweenness variables to be binary.

The reader may have noticed that this MILO model does not account for the vertical component of the distance between departments. This can be done by introducing y variables and including the appropriate constraints based on the modelling in Sect. 3.2.2.

3.4 Multi-Row Facility Layout with Departments of Equal Length

This section considers the special case of the MRFLP in which the departments all have the same length. This means that we can set the lengths of all the departments equal to 1 without loss of generality. We also assume here that the set $R = \{1, 2, \ldots, m\}$ of rows available for placing the departments is given. This special case is called the multi-row equidistant facility layout problem (MREFLP) and is also known as the equidistant MRFLP. We point out that the formulations in this section can be specialized to the double-row case (DREFLP).

Before stating the formulations for the MREFLP, we present in Sect. 3.4.1 some theoretical results that derive from its special structure, namely that all the department lengths are equal. We use these results to write the formulations in the subsequent sections.

3.4.1 Properties of Optimal Solutions for the MREFLP

A fundamental observation is the fact that because all the departments have unit length, we can define the notion of *columns* within the row layout, as illustrated in Fig. 3.2 where departments 1 and 2 are in the same column, as are departments 3 and 5.

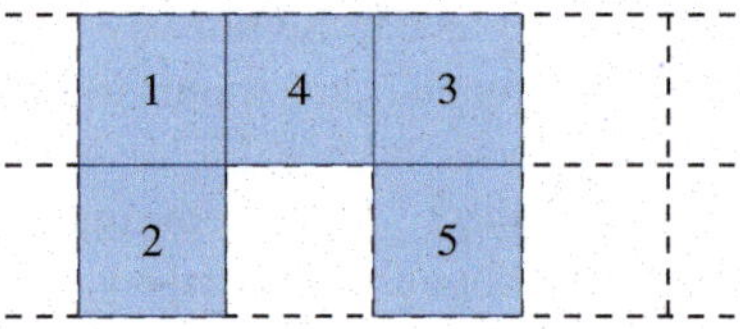

Fig. 3.2 Columns in the MREFLP

Specifically, we have the following result.

Theorem 3.2 *There is always an optimal solution to the MREFLP on the integer grid.*

Theorem 3.2 is especially interesting because it allows us to make specific statements about the empty spaces that can occur between departments in the same row, which is the last of the three tasks stated at the beginning of this chapter. In particular, it implies that all the spaces between departments will have integer lengths.

Moreover, with a careful and detailed analysis, it is possible to make specific statements about the minimum number of columns required to obtain at least one optimal layout for the MREFLP. First, we make the following assumptions, where n is the number of departments, and m is the number of rows available for the layout.

Assumption 1 Columns containing only spaces can be deleted.

Assumption 2 If two nonempty neighbouring columns together contain no more than m departments, then all the corresponding departments can be assigned to the left column, and the right column can be deleted.

Assumption 3 If $n > 2m$ and the first and third columns contain in total at most m departments, then all the corresponding departments can be assigned to the third column, and the first column can be deleted. More generally, this holds for columns $k' - 2$ and k' such that each column with index at most k' contains at least one department.

Under these assumptions, the following theorem can be proved.

Theorem 3.3 *The minimum number of columns sufficient to preserve at least one optimal layout for an instance with n departments is*

1. *equal to 1 if $n \leq m$ and equal to 2 if $m < n < \frac{3}{2}m + \frac{3}{2}$;*
2. *equal to $\left\lceil \frac{2n}{3} \right\rceil - 1$ for the DREFLP with $n \geq 9$;*
3. *equal to $\left\lfloor \frac{2n}{m+1} \right\rfloor$ for the MREFLP with an odd number of rows m; and*
4. *equal to or at most $2t + 1$ for the MREFLP with an even number of rows m and $n \in \{\frac{m}{2} + 2 + (m+1)(t-1), \ldots, \frac{m}{2} + 1 + (m+1)t\}$ for some $t \in \mathbb{N}$.*

This means that we can compute in advance the number of columns required to obtain an optimal solution. We call this number C. Knowing C and knowing that we have n departments, we can fill up the C columns and m rows with spacing departments, i.e., departments of length 1 and with all pairwise connectivities involving them equal to 0. This means that there is no need for models to explicitly consider the issue of empty spaces between departments in the same row, as all the necessary spaces will be occupied by the spacing departments in an optimal way.

3.4.2   Integer Linear Optimization Model

In this section we present an integer linear optimization (ILO) model for the MREFLP. The approach used here takes advantage of the structure induced by the fact that all the department lengths are equal, and the result is a formulation in which all the variables are binary. In fact, much stronger results can be proved about the MREFLP, and they allow us to build much tighter models than those for the general MRFLP.

The information about the column-wise relative position of pairs i, j of departments is incorporated into the model using the following variables:

$$\xi_{ij} = \begin{cases} 1, & \text{if departments } i \text{ and } j \text{ are assigned to the same column} \\ 0, & \text{otherwise.} \end{cases}$$

We also recall again the betweenness variables from Sect. 2.4. For three distinct departments i, j, and k, we have

$$\beta_{ijk} = \begin{cases} 1, & \text{if department } k \text{ lies between departments } i \text{ and } j \\ 0, & \text{otherwise.} \end{cases}$$

Note that the definition of β_{ijk} counts all departments k between i and j regardless of which row k is in. Recall that by Theorem 3.3, we can fill up the C columns and m rows with spacing departments. This means that for each column between i and j, there are precisely m departments counted as being between i and j.

We can now formulate the MREFLP as an ILO problem:

$$\text{minimize} \quad \sum_{i<j} c_{ij} \left(\frac{1}{m} \sum_{k \neq i,j} \beta_{ijk} + (1 - \xi_{ij}) \right) \tag{3.66}$$

$$\text{s.t.} \quad \beta_{ijh} + \beta_{ikh} + \beta_{jkh} \leq 2, \quad i < k < j, \ h \neq i, j, k, \tag{3.67}$$

$$-\beta_{ijh} + \beta_{ikh} + \beta_{jkh} + \beta_{ijk} \geq 0, \quad i < k < j, \ h \neq i, j, k, \tag{3.68}$$

$$\beta_{ijh} - \beta_{ikh} + \beta_{jkh} + \beta_{ikj} \geq 0, \quad i < k < j, \ h \neq i, j, k, \tag{3.69}$$

$$\beta_{ijh} + \beta_{ikh} - \beta_{jkh} + \beta_{jki} \geq 0, \quad i < k < j, \ h \neq i, j, k, \tag{3.70}$$

$$\xi_{hk} - \beta_{ijh} + \beta_{ikh} + \beta_{jkh} \geq 0, \quad i < k < j, \ h \neq i, j, k, \tag{3.71}$$

$$\xi_{hj} + \beta_{ijh} - \beta_{ikh} + \beta_{jkh} \geq 0, \quad i < k < j, \ h \neq i, j, k, \tag{3.72}$$

$$\xi_{hi} + \beta_{ijh} + \beta_{ikh} - \beta_{jkh} \geq 0, \quad i < k < j, \ h \neq i, j, k, \tag{3.73}$$

$$\xi_{ij} + \beta_{ikj} + \beta_{ijk} + \beta_{jki} \leq 1, \quad i < k < j, \tag{3.74}$$

$$\xi_{ik} + \beta_{ikj} + \beta_{ijk} + \beta_{jki} \leq 1, \quad i < k < j, \tag{3.75}$$

$$\xi_{jk} + \beta_{ikj} + \beta_{ijk} + \beta_{jki} \leq 1, \quad i < k < j, \tag{3.76}$$

$$\xi_{ij} + \xi_{jk} - \xi_{ik} \leq 1, \quad i < j, \ k \neq i, j, \tag{3.77}$$

$$\sum_{\substack{j \neq i}} \xi_{ij} = m - 1, \quad 1 \leq i \leq n, \tag{3.78}$$

$$\sum_{i=1}^{n} \sum_{k=i+1}^{n} \sum_{\substack{j=1 \\ j \neq i,k}}^{n} \beta_{ijk} = m^3 \binom{C}{3}, \tag{3.79}$$

$$\beta_{ijk}, \xi_{ij} \in \{0, 1\}, \quad 1 \leq i < j \leq n, \ k \neq i, j \tag{3.80}$$

where, as above, C is the number of columns fixed according to Theorem 3.3, and m is the number of rows.

The objective function (3.66) expresses the distance between i and j by counting the number of departments between i and j. Because it counts all the departments between i and j over all rows, this total must be divided by m. Furthermore, if i and j are not in the same column ($\xi_{ij} = 0$), then it increases the distance by 1 (which is half of the sum of the lengths of i and j).

Constraints (3.67) ensure that, for every choice of three distinct departments i, j, and k, department h cannot simultaneously lie between i and j, between i and k, and between j and k.

Constraints (3.68)–(3.73) require a form of consistency, namely that for every choice of three distinct departments i, j, and k, if department h lies between departments i and j, then also h lies between i and k, or between j and k, or in the same column as k (which also implies that k lies between i and j).

Constraints (3.74)–(3.76) state that if two departments out of three are in the same column, then none of them is between the other two. Simultaneously, if one of the three departments is between the other two, then no two of them are in the same column. Constraints (3.77) are transitivity constraints. Constraints (3.78) say that each department shares its column with $m - 1$ other departments. Finally, constraint (3.79) counts the number of betweenness variables that must equal 1. The idea is that for each of the $\binom{C}{3}$ subsets of three columns, there are m^3 ways to choose one department from each column, and these are the only choices for which the variable β_{ijk} equals 1.

This formulation can be specialized for the double-row case. For the objective function, we set $m = 2$. Constraints (3.67)–(3.73) are still valid, and constraints (3.74)–(3.76) can be strengthened to

$$\xi_{ij} + \xi_{ik} + \xi_{jk} + \beta_{ijk} + \beta_{ikj} + \beta_{jki} = 1, \quad i < k < j.$$

Constraints (3.77) are not changed, and for constraints (3.78) and (3.79), we set $m = 2$.

3.5 Additional Special Cases of Multi-Row Facility Layout

In this section we present other special cases for layout on several rows with specific additional structure.

3.5.1 Corridor Allocation Problem

The corridor allocation problem (CAP), like the DRFLP, is concerned with assigning departments to two rows. However, the CAP has the following additional conditions:

- no empty space is allowed between two adjacent departments;
- the leftmost points of each row must be aligned, typically at the origin.

Note that because the CAP is a version of the DRFLP with additional restrictions, it follows that for the same set of departments and connectivities, the optimal cost of the CAP will always be at least as high as that of the DRFLP. This is illustrated in Example 3.2 where the optimal solutions of the DRFLP and CAP are different for the same set of departments with the same connectivities.

Example 3.2 We consider an instance with 5 departments. The length of departments $1, 2$, and 3 is 8, and the length of departments 4 and 5 is 2. The nonzero connectivities are $c_{12} = c_{45} = 1$ and $c_{13} = c_{24} = 10$. The optimal solutions for the DRFLP and CAP are illustrated below (Figs. 3.3 and 3.4).

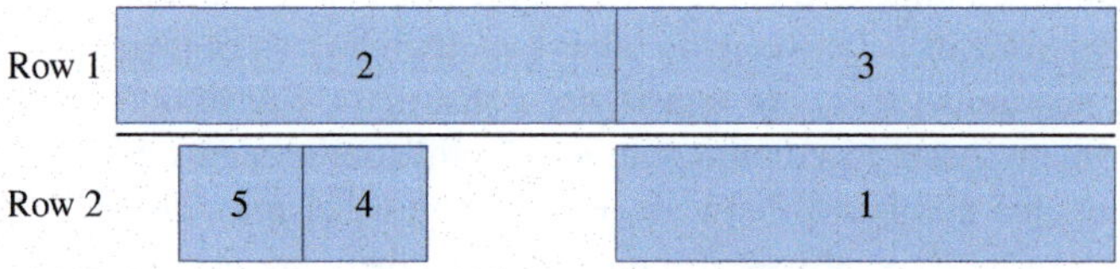

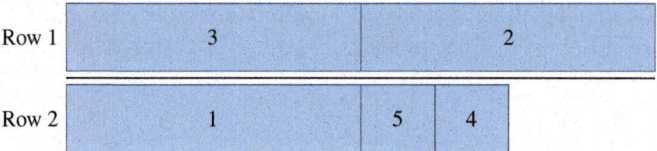

Fig. 3.3 Optimal layout for the DRFLP with a total cost equal to 10

Fig. 3.4 Optimal layout for the CAP with a total cost equal to 20

The CAP can be formulated in a similar way to the DRFLP. The formulation that we give below is based on the formulation of the DRFLP in Sect. 3.1.1.

$$\text{minimize} \quad \sum_{i=1}^{n-1} \sum_{j=i+1}^{n} c_{ij} d_{ij} \tag{3.81}$$

$$\text{s.t.} \quad d_{ij} \geq x_i - x_j, \ d_{ij} \geq x_j - x_i, \quad 1 \leq i < j \leq n, \tag{3.82}$$

$$x_i = \frac{\ell_i}{2} + \sum_{k \neq i} \ell_k \alpha_{ki}, \quad 1 \leq i \leq n, \tag{3.83}$$

$$d_{ij} - \left(\frac{\ell_i + \ell_j}{2} \right) \alpha_{ij} - \left(\frac{\ell_i + \ell_j}{2} \right) \alpha_{ji} \geq 0, \quad 1 \leq i < j \leq n \tag{3.84}$$

$$(3.1) - (3.4)$$

$$\alpha_{ij} \in \{0, 1\}, \quad 1 \leq i \neq j \leq n, \tag{3.85}$$

$$\frac{\ell_i}{2} \leq x_i \leq L - \ell_i, \quad 1 \leq i \leq n. \tag{3.86}$$

The main difference is the way of expressing the position of each department within its row, because for the CAP, the position x_i of department i is exactly equal to the sum of the lengths of the preceding departments (see constraints (3.83)). The formulation can be simplified by using (3.83) to substitute for x_i and x_j in (3.82) and (3.86). We can then omit constraints (3.83) and (3.84), and the distances are computed by the new form of constraints (3.82):

$$d_{ij} \geq \frac{\ell_i - \ell_j}{2} + \sum_{k \neq i} \ell_k \alpha_{ki} - \sum_{k \neq j} \ell_k \alpha_{kj}, \quad 1 \leq i < j \leq n, \tag{3.87}$$

$$d_{ij} \geq \frac{\ell_j - \ell_i}{2} + \sum_{k \neq j} \ell_k \alpha_{kj} - \sum_{k \neq i} \ell_k \alpha_{ki}, \quad 1 \leq i < j \leq n. \tag{3.88}$$

3.5.2   k-Corridor Allocation Problem

The k-corridor allocation problem (k-CAP) is the generalization of the CAP (where $k = 2$) to three or more rows. It is also known as the space-free multi-row facility layout problem (SF-MRFLP). Solutions to the k-CAP must satisfy the conditions of the CAP (no space between adjacent departments; leftmost point of every row at origin). The k-CAP can be approached by solving an instance of the k-PROP (see Sect. 3.5.4) for every possible assignment of departments to rows. However, this strategy can solve only small instances.

3.5.3   Parallel Row Ordering Problem

The parallel row ordering problem (PROP) is a version of the CAP in which, in addition, the assignment of departments to rows is given. Let m be the number of departments and N_1 and N_2 be a partition of the set of departments $\{1, 2, \ldots, m\}$, i.e., $N = N_1 \cup N_2$ and $N_1 \cap N_2 = \emptyset$ so that the sets N_1 and N_2 represent an assignment of the departments to the two rows. To obtain a formulation for the PROP, it suffices to add the following constraints to the CAP (3.81)–(3.86):

$$\alpha_{ij} = \alpha_{ji} = 0, \quad i \in N_1,\ j \in N_2$$

$$\alpha_{ij} + \alpha_{ji} = 1, \quad i, j \in N_1 \text{ or } i, j \in N_2.$$

3.5.4   k-Parallel Row Ordering Problem

The k-parallel row ordering problem (k-PROP) is the generalization of the PROP (where $k = 2$) to three or more rows. As for the PROP, the assignment of departments to rows is given, and solutions to the k-PROP must satisfy the conditions of the k-CAP (no space between adjacent departments; leftmost point of every row at origin).

The MILO formulation described here is a modification of the model presented in Sect. 3.3 for the FR-MRFLP. We use the notation of Sect. 3.3, so that $R = \{1, 2, \ldots, m\}$ denotes the set of rows and N_r the set of departments assigned to row $r \in R$. We also use the dummy departments $n + 1$ and $n + 2$ to be placed at the left and right boundaries, respectively, of the layout, and $\ell_{n+1} = \ell_{n+2} = 0$, as well as $c_{ij} = 0$ if $i \in \{n + 1, n + 2\}$ or $j \in \{n + 1, n + 2\}$. Also as before, we extend the set N_r to include the dummy departments: $\tilde{N}_r = N_r \cup \{n + 1, n + 2\}$.

The modifications made here to the MILO model from Sect. 3.3 guarantee the two k-CAP conditions. Specifically, we change the constraints defining the distance variables. The new model is

$$\text{minimize} \quad \sum_{\substack{i, j \in N \\ i < j}} c_{ij} d_{ij} \tag{3.89}$$

$$\text{s.t.} \quad (3.52) - (3.60), \quad (3.65)$$

$$d_{ij} = \sum_{\substack{k \in N_r \\ k \neq i, j}} \ell_k \beta_{ijk} + \frac{\ell_i + \ell_j}{2}, \quad r \in R,\ i, j \in N_r \cup \{n + 1\},\ i < j, \tag{3.90}$$

$$d_{ij} \geq 0, \quad i, j \in N \cup \{n + 1\},\ i < j. \tag{3.91}$$

Constraints (3.90) define the distance between each pair of departments in the same row. Because the distance between departments i and j is the sum of the lengths of all the departments between them, there will be no empty space between adjacent departments, including the dummy departments.

3.6 Semidefinite Optimization Formulations

Semidefinite optimization models have been proposed for the DRFLP and for several of the variants of the MRFLP presented in this chapter. These models apply the principles of Sect. 2.7 to derive an SDO approach, but the technical details are much more involved. Moreover, in most cases, specific solution algorithms must be implemented to achieve satisfactory performance with the SDO approach. For this reason, we merely briefly sketch the SDO approach for the k-PROP, and how it can be applied to instances of the DRFLP and MRFLP. The references provided in Sect. 3.7 provide all the details.

Consider the k-PROP with n departments and m rows, and let the assignment of departments to rows be specified by the mapping $r : \{1, \ldots, n\} \to \{1, \ldots, m\}$. Define the binary variables γ_{ij} as in Sect. 2.7, and let d_{ij} represent the centre-to-centre distance between i and j measured parallel to the rows.

If i and j are assigned to the same row, i.e., if $r(i) = r(j)$, then

$$d_{ij} = \frac{1}{2}(\ell_i + \ell_j) + \sum_{\substack{k \in N, \, k < i \\ r(k) = r(i)}} \ell_k \frac{1 - \gamma_{ki}\gamma_{kj}}{2} + \sum_{\substack{k \in N, \, i < k < j \\ r(k) = r(i)}} \ell_k \frac{1 + \gamma_{ik}\gamma_{kj}}{2}$$
$$+ \sum_{\substack{k \in N, \, k > j \\ r(k) = r(i)}} \ell_k \frac{1 - \gamma_{ik}\gamma_{jk}}{2}, \tag{3.92}$$

and if $r(i) \neq r(j)$,

$$d_{ij} = \gamma_{ij} \left[\left(\frac{\ell_j}{2} + \sum_{\substack{k \in N, \, k < j \\ r(k) = r(j)}} \ell_k \frac{1 + \gamma_{kj}}{2} + \sum_{\substack{k \in N, \, k > j \\ r(k) = r(j)}} \ell_k \frac{1 - \gamma_{jk}}{2} \right) - \left(\frac{\ell_i}{2} + \sum_{\substack{k \in N, \, k < i \\ r(k) = r(i)}} \ell_k \frac{1 + \gamma_{ki}}{2} + \sum_{\substack{k \in N, \, k > i \\ r(k) = r(i)}} \ell_k \frac{1 - \gamma_{ik}}{2} \right) \right]. \tag{3.93}$$

The above relations, plus the triangle inequalities for the distances between every triplet of departments i, j, and k

$$d_{ij} + d_{ik} \geq d_{jk}, \quad d_{ij} + d_{jk} \geq d_{ik}, \quad d_{ik} + d_{jk} \geq d_{ij}, \quad 1 \leq i < j < k \leq n, \tag{3.94}$$

can be used to obtain an SDO formulation for the k-PROP.

In the SDO formulation, the possibility of spaces is handled using the next theorem. The result makes use of the half-integer grid, which is the set of all values that are equal to an integer or an integer plus 0.5. For example, 5 and 9.5 are both half-integer, but 4.3 is not.

Theorem 3.4 *If all the department lengths ℓ_i are integer, then there is always an optimal solution to the MRFLP on the half-integer grid.*

Corollary 3.2 *If all the department lengths ℓ_i are integer, then for each instance of the MRFLP, we obtain an equivalent instance of the k-PROP by adding spacing departments of length 0.5 such that the length of each row becomes equal to $M := \sum_{i=1}^{n} \ell_i$.*

We can solve a DRFLP or a MRFLP using the k-PROP SDO-based formulation by adding enough spacing departments of length 0.5 with all involved connectivities equal to zero and then applying the SDO approach for k-PROP. Because the number of spacing departments needed will normally be too large for computation, it is in practice necessary to reduce this number.

Finally, the restriction that the assignment of departments to rows is fixed can be handled by using the above approach to find the global optimal solution for each of the possible assignments (or for a subset of them). This will lead to a global optimal solution (or global lower bounds) for the DRFLP or the MRFLP.

3.7 References and Further Reading

The relationship between the material handling system and the type of layout is discussed in Heragu and Kusiak (1988) and Heragu (2008). Another application of the DRFLP is the arrangement of rooms in buildings, see, e.g., Ahonen et al (2014).

The MILO formulation for the DRFLP in Sect. 3.1.1 was proposed by Amaral (2013). Subsequently, Secchin and Amaral (2019) presented the model in Sect. 3.1.2, but we have presented the nonoverlap constraint (3.13) instead of

$$x_i + d_{ij} \leq x_j + 2(L - \ell_i/2 - \ell_j/2)(1 - \alpha_{ij}), \quad i < j$$
$$x_i + d_{ji} \leq x_j + 2(L - \ell_i/2 - \ell_j/2)(1 - \alpha_{ij}), \quad i > j.$$

When i and j are set to the same row, these constraints together with (3.12) make $d_{ij} = |x_j - x_i|$. Since the advantage of these constraints is unclear, we retained the nonoverlap constraints (3.13), which are a common feature of all layout problems.

The MRFLP has received fairly limited attention in the operations research literature to date. The MILO formulation for MRFLP in Sect. 3.2.1 was proposed in Chung and Tanchoco (2010); their model also explicitly accounts for clearances between departments. This model was later corrected in Zhang and Murray (2012). The formulation in Sect. 3.2.2 was introduced in Anjos and Vieira (2021), where Theorem 3.1 and Corollary 3.1 are proved. Total unimodularity is a classical topic in integer optimization. The results quoted in Sect. 3.2.3 are from Nemhauser and Wolsey (1988, Section III.2), where the reader can find a wealth of information about TU matrices and their applications.

The theoretical results and the ILO formulation for the MREFLP given in Sect. 3.4 were presented in Anjos et al (2018), which also discusses their specialization for the DRFLP. The MILO formulation for the CAP presented in Sect. 3.5.1 was introduced in Amaral (2012).

The models for k-PROP in Sect. 3.5.4 and for the FR-MRFLP in Sect. 3.3 were introduced in Fischer et al (2019). They presented these models in the opposite order: first, a model for the k-PROP is introduced, and then the FR-MRFLP is considered. Together with these models, Fischer et al (2019) present an algorithm (based on the enumeration of row assignments) that allows them to solve the MRFLP and k-CAP. They develop strategies for reducing the number of possible row assignments. As of the time of writing, this is the most efficient strategy for instances of the MRFLP and k-CAP.

The SDO model, the theoretical results, and the k-PROP approach for the MRFLP in Sect. 3.6 were presented in Hungerländer and Anjos (2015). They prove several results to reduce the number of spacing departments needed and hence improve the computational performance of the SDO approach.

References

Ahonen H, de Alvarenga A, Amaral A (2014) Simulated annealing and tabu search approaches for the corridor allocation problem. Eur J Oper Res 232(1):221–233

Amaral ARS (2012) The corridor allocation problem. Comput Oper Res 39(12):3325–3330

Amaral ARS (2013) Optimal solutions for the double row layout problem. Optimization Letters 7(2):407–413

Anjos MF, Vieira MVC (2021) Mathematical optimization approach for facility layout on several rows. Optimization Letters 15:9–23

Anjos MF, Fischer A, Hungerländer P (2018) Improved exact approaches for row layout problems with departments of equal length. Eur J Oper Res 270(2):514–529

Chung J, Tanchoco JMA (2010) The double row layout problem. Int J Prod Res 48(3):709–727

Fischer A, Fischer F, Hungerländer P (2019) New exact approaches to row layout problems. Math Programm Comput 11:703–754

Heragu SS (2008) Facilities design. CRC Press

Heragu SS, Kusiak A (1988) Machine layout problem in flexible manufacturing systems. Operations Research 36(2):258–268

Hungerländer P, Anjos MF (2015) A semidefinite optimization-based approach for global optimization of multi-row facility layout. Eur J Oper Res 245(1):46–61

Nemhauser GL, Wolsey LA (1988) Integer and combinatorial optimization. Wiley-Interscience, New York, NY, USA

Secchin LD, Amaral ARS (2019) An improved mixed-integer programming model for the double row layout of facilities. Optimization Letters 13(1):193–199

Zhang Z, Murray CC (2012) A corrected formulation for the double row layout problem. Int J Prod Res 50(15):4220–4223

Chapter 4
Layout of a Single Floor

This chapter is concerned with the problem of physically arranging a set of two-dimensional departments inside a two-dimensional facility. Mathematically, the problem is to find the optimal placement of a given number of nonoverlapping indivisible departments, with a given area requirement for each department, so as to minimize the total cost of flows inside the floor. This single floor facility layout problem is often called the unequal-areas facility layout problem (UA-FLP), and it differs substantially from row layouts. The departments are now genuinely two-dimensional, and determining the optimal dimensions for each department such that the area requirements are met is part of the problem.

4.1 Nonconvex Continuous Optimization Formulation

We begin with an exact formulation of the UA-FLP using only continuous variables. This formulation is of limited use in practice, but it is straightforward, it allows us to establish the notation for this chapter, and it highlights the challenging aspects of the UA-FLP, thus motivating the solution approaches presented in the remainder of the chapter.

The constraints for the UA-FLP can be grouped into two sets:

- *Department shape constraints* ensure the required area of each department and enforce restrictions on its dimensions (height and width). These requirements generally lead to linear or convex constraints but still pose some challenges.
- *Department location constraints* place every department within the facility and ensure that there is no overlap between pairs of departments. The main challenge here is the nonoverlap constraints that are inherently nonconvex and combinatorial.

© Springer Nature Switzerland AG 2021
M. F. Anjos, M. V. C. Vieira, *Facility Layout*, EURO Advanced Tutorials
on Operational Research, https://doi.org/10.1007/978-3-030-70990-7_4

To model the UA-FLP, we use four continuous variables per department, namely (x_i, y_i) to represent the coordinates of the centre of department i and h_i and w_i to represent its height and width. The two-dimensional reference system for the coordinates (x_i, y_i) has its origin at the centre of the facility. We assume that the required area A_i for each department i is given. We also assume that we are given lower and upper bounds $h_i^{\min}$ and $h_i^{\max}$ on the height of department i and $w_i^{\min}$ and $w_i^{\max}$ on its width.

Moreover, the dimensions h_i and w_i of department i are normally required to be balanced. To achieve this, we set an upper bound on the *aspect ratio* of each department, which is defined as the larger of the two ratios height/width and width/height. We let ρ_i be a given upper bound on the aspect ratio of department i. Clearly, $\rho_i \geq 1$ must hold, and the closer ρ_i is to unity, the closer the shape of department i will be to a square. Requiring the aspect ratio to be small is desirable in real-world applications, but it generally makes the problem harder.

Finally, we let h_F and w_F denote, respectively, the height and width of the facility. We assume throughout this chapter that these are a given input, but more generally they can be optimized in the same way that we optimize the dimensions of each department.

With this notation, our first formulation of the UA-FLP is as follows:

$$\text{minimize} \quad \sum_{1 \leq i < j \leq n} c_{ij}(|x_i - x_j| + |y_i - y_j|) \tag{4.1}$$

$$\text{s.t.} \quad h_i^{\min} \leq h_i \leq h_i^{\max}, \quad 1 \leq i \leq n, \tag{4.2}$$

$$w_i^{\min} \leq w_i \leq w_i^{\max}, \, 1 \leq i \leq n, \tag{4.3}$$

$$w_i h_i = A_i, \quad 1 \leq i \leq n \tag{4.4}$$

$$\max\left\{\frac{w_i}{h_i}, \frac{h_i}{w_i}\right\} \leq \rho_i, \quad 1 \leq i \leq n, \tag{4.5}$$

$$x_i + \frac{1}{2}w_i \leq \frac{1}{2}w_F, \quad \frac{1}{2}w_i - x_i \leq \frac{1}{2}w_F, \quad 1 \leq i \leq n, \tag{4.6}$$

$$y_i + \frac{1}{2}h_i \leq \frac{1}{2}h_F, \quad \frac{1}{2}h_i - y_i \leq \frac{1}{2}h_F, \quad 1 \leq i \leq n, \tag{4.7}$$

$$\begin{aligned} |x_i - x_j| &\geq \tfrac{1}{2}(w_i + w_j) \quad \textbf{or} \\ |y_i - y_j| &\geq \tfrac{1}{2}(h_i + h_j), \quad 1 \leq i < j \leq n. \end{aligned} \tag{4.8}$$

We look at each part of this formulation in turn. The objective function (4.1) is not linear, but its absolute value terms can be linearized as demonstrated in Sect. 2.3.1.

The first four sets of constraints enforce the shape requirements. Constraints (4.2) and (4.3) enforce the lower and upper bounds on the height and width of department i. Constraints (4.4) ensure that each department has the prescribed area, and constraints (4.5) enforce the maximum aspect ratio for each department.

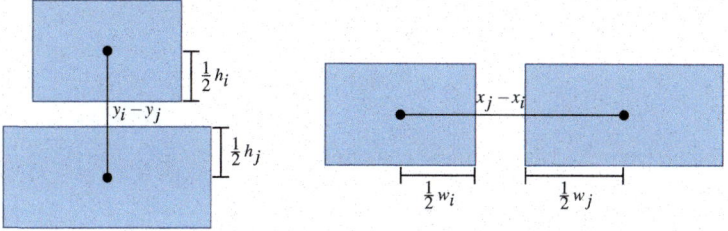

Fig. 4.1 Diagram showing that the nonoverlap constraints must ensure that either $|y_i - y_j| \geq \frac{1}{2}(h_i + h_j)$ or $|x_i - x_j| \geq \frac{1}{2}(w_i + w_j)$

The last three sets of constraints enforce the location requirements. Constraints (4.6) and (4.7) ensure that the departments are placed entirely inside the facility. Constraints (4.8) prevent overlapping by forcing the separation of each pair of departments in (at least) one direction (see Fig. 4.1).

The area constraints (4.4) are bilinear (not linear), and they have been tradition-ally modelled using linearization approaches (see Sect. 4.6). Alternatively, they can be handled directly using conic optimization. Both approaches rely on the observation that these constraints can be relaxed to

$$w_i h_i \geq A_i, \tag{4.9}$$

which has the advantage of being convex. Moreover, this relaxed form can be formulated as a semidefinite constraint,

$$\begin{bmatrix} h_i & \sqrt{A_i} \\ \sqrt{A_i} & w_i \end{bmatrix} \succeq 0 \tag{4.10}$$

or, equivalently, as a second-order cone constraint (see Appendix A),

$$w_i + h_i \geq \left\| \left(\begin{array}{c} w_i - h_i \\ 2\sqrt{A_i} \end{array} \right) \right\|_2 . \tag{4.11}$$

Because the optimization will push this relaxed form towards equality, in the optimal solution $w_i h_i$ will be close to or equal to A_i. Moreover, it is straightforward to check that if $\sum_{i=1}^{n} A_i = h_F w_F$, then any one of the relaxed forms (4.9), (4.10), and (4.11) suffices to ensure that (4.4) holds at every feasible solution.

Constraints (4.5) enforce the maximum aspect ratio; it is straightforward to write each of them as two linear inequality constraints:

$$\max \left\{ \frac{w_i}{h_i}, \frac{h_i}{w_i} \right\} \leq \rho_i$$

is equivalent to

$$\frac{w_i}{h_i} \leq \rho_i \quad \text{and} \quad \frac{h_i}{w_i} \leq \rho_i,$$

which can be expressed as the linear constraints

$$w_i \leq \rho_i h_i \quad \text{and} \quad h_i \leq \rho_i w_i.$$

The nonoverlap constraints (4.8) are disjunctive and nonconvex by their very nature, and they are the hardest to enforce. Much of the rest of this chapter is concerned with ways to handle these constraints. If the relative position of each pair of departments is known, then constraints (4.8) can be written as linear inequalities, and the formulation becomes a convex optimization problem that is straightforward to solve. This observation motivates the two-stage approaches presented in Sect. 4.4.

We recall that this formulation locates the centre of the facility at the origin, while other models locate the bottom left-hand corner at the origin. This difference changes the mathematical details slightly but is otherwise of no consequence.

4.2   Mixed-Integer Second-Order Conic Optimization Formulation

We present in this section a mixed-integer second-order conic optimization (MIS-OCO) formulation of the UA-FLP that uses binary variables to linearize the nonoverlap constraints (4.8). Specifically, we define four binary variables for each pair of departments. For all $1 \leq i \leq n$ and $1 \leq j \leq n$, let

$$\alpha_{ij} = \begin{cases} 1 & \text{if } i \text{ is to the left of } j, \text{ i.e., } i \text{ precedes } j \text{ in the horizontal dimension} \\ 0 & \text{otherwise,} \end{cases}$$

$$\beta_{ij} = \begin{cases} 1 & \text{if } i \text{ is below } j, \text{ i.e., } i \text{ precedes } j \text{ in the vertical dimension} \\ 0 & \text{otherwise.} \end{cases}$$

These definitions are illustrated in Fig. 4.2.

Fig. 4.2 We set $\alpha_{ij} = 1$ when department i is to the left of j, and $\beta_{ij} = 1$ when department i is below j

Furthermore, recall that d_{ij}^x and d_{ij}^y represent the horizontal and vertical distances between i and j. Using these variables, we obtain the following MISOCO formulation:

$$\text{minimize} \quad \sum_{1 \le i < j \le n} c_{ij}(d_{ij}^x + d_{ij}^y) \tag{4.12}$$

$$\text{s.t.} \quad d_{ij}^x \ge x_i - x_j, \quad d_{ij}^x \ge x_j - x_i, \quad 1 \le i < j \le n, \tag{4.13}$$

$$d_{ij}^y \ge y_i - y_j, \quad d_{ij}^y \ge y_j - y_i, \quad 1 \le i < j \le n, \tag{4.14}$$

$$h_i^{\min} \le h_i \le h_i^{\max}, \quad 1 \le i \le n, \tag{4.15}$$

$$w_i^{\min} \le w_i \le w_i^{\max}, \quad 1 \le i \le n, \tag{4.16}$$

$$w_i + h_i \ge \left\| \begin{pmatrix} w_i - h_i \\ 2\sqrt{A_i} \end{pmatrix} \right\|_2, \quad 1 \le i \le n \tag{4.17}$$

$$w_i \le \rho_i h_i \quad \text{and} \quad h_i \le \rho_i w_i, \quad 1 \le i \le n, \tag{4.18}$$

$$\frac{1}{2}(w_i - w_F) \le x_i \le \frac{1}{2}(w_F - w_i), \quad 1 \le i < j \le n, \tag{4.19}$$

$$\frac{1}{2}(h_i - h_F) \le y_i \le \frac{1}{2}(h_F - h_i), \quad 1 \le i < j \le n, \tag{4.20}$$

$$x_i + \frac{1}{2}w_i \le x_j - \frac{1}{2}w_j + w_F(1 - \alpha_{ij}), \quad 1 \le i \ne j \le n, \tag{4.21}$$

$$y_i + \frac{1}{2}h_i \le y_j - \frac{1}{2}h_j + h_F(1 - \beta_{ij}), \quad 1 \le i \ne j \le n, \tag{4.22}$$

$$\alpha_{ij} + \alpha_{ji} + \beta_{ij} + \beta_{ji} = 1, \quad 1 \le i < j \le n, \tag{4.23}$$

$$\alpha_{ij}, \beta_{ij} \in \{0, 1\}, \quad 1 \le i, j \le n. \tag{4.24}$$

Several constraints in this formulation are similar to those in the formulation of Sect. 4.1. The objective function (4.12) and constraints (4.13) and (4.14) provide a linearization of the objective function (4.1) by applying the first linearization approach described in Sect. 2.3.1 to each of the x and y directions. Constraints (4.15) and (4.16) are unchanged and enforce the lower and upper bounds on the height and width of department i. Constraints (4.17) are the relaxed form of the area constraints (4.4), and constraints (4.18) are the unchanged aspect ratio constraints. Constraints (4.19) and (4.20) are slightly rewritten forms of (4.6) and (4.7).

The last four sets of constraints are the new nonoverlap constraints. Constraints (4.23) require that for each pair i, j of departments, precisely one of the four corresponding binary variables must be equal to 1 (and the other three must be 0). Constraints (4.21) and (4.22) are linearized versions of (4.8) that enforce the nonoverlap requirement for i and j corresponding to the binary variable set to 1. Finally, constraints (4.24) require α_{ij} and β_{ij} to be binary.

Let us look more closely at how this formulation of the nonoverlap constraints works. For example, consider departments $i = 1$ and $j = 2$, and suppose that $\alpha_{12} = 1$. By constraint (4.23), we have that $\alpha_{21} = 0$, $\beta_{12} = 0$, and $\beta_{21} = 0$. Constraints (4.21) become

$$x_1 + \frac{1}{2}w_1 \leq x_2 - \frac{1}{2}w_2 \quad \text{and} \quad x_2 + \frac{1}{2}w_2 \leq x_1 - \frac{1}{2}w_1 + w_F,$$

and constraints (4.22) become

$$y_1 + \frac{1}{2}h_1 \leq y_2 - \frac{1}{2}h_2 + h_F \quad \text{and} \quad y_2 + \frac{1}{2}h_2 \leq y_1 - \frac{1}{2}h_1 + h_F.$$

The first constraint requires that the left border of department 2 (with x coordinate equal to $x_2 - \frac{1}{2}w_2$) must lie to the right of the right border of department 1 (with x coordinate equal to $x_1 + \frac{1}{2}w_1$). In other words, department 1 is to the left of 2, or equivalently, 1 precedes 2 in the horizontal dimension, as desired from the choice of $\alpha_{12} = 1$. Moreover, the other three constraints have the width or height of the entire facility (w_F or h_F) added to their right-hand side. This renders them inactive because they will be satisfied by any otherwise feasible values of the variables involved. That is, w_F and h_F are effectively acting as big-M values; see Sect. 2.3.3.

An alternative MISOCO representation of the relative positions of pairs of departments is introduced in the next section.

4.3 Sequence-Pair Formulation

Sequence-pair approaches determine the relative positions of the departments using the so-called *sequence-pair representation*. This representation is integrated into a MISOCO model similar to that in Sect. 4.2 to obtain a formulation of the UA-FLP.

A sequence-pair, denoted Γ_+ and Γ_-, consists of a pair of sequences of the set of departments. Together, these two sequences encode the relative position of each pair of departments.

The following definition explains how a sequence-pair defines the relative positions of departments within a layout.

Definition 4.1 Let (Γ_+, Γ_-) be a sequence-pair, and let i and j be two departments in Γ_+, Γ_-. The sequence-pair determines the relative horizontal or vertical placement of i and j in a layout as follows:

- if i follows j in both Γ_+ and Γ_-, then i is to the right of j;
- if i precedes j in both Γ_+ and Γ_-, then i is to the left of j;
- if i precedes j in Γ_+ and follows j in Γ_-, then i is above j;
- if i follows j in Γ_+ and precedes j in Γ_-, then i is below j.

Fig. 4.3 Layout
corresponding to the
sequence-pair
$\Gamma_+ = \{415263\}$ and
$\Gamma_- = \{124536\}$

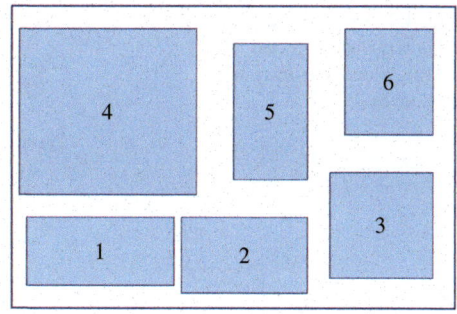

Figure 4.3 shows an example of a layout of six departments for which a sequence-pair is

$$\Gamma_+ = \{415263\}, \quad \Gamma_- = \{124536\}.$$

We now explain how the sequence-pair structure can be incorporated into a MISOCO model. Given a sequence-pair (Γ_+, Γ_-), we define the following binary variables for all $1 \leq i \leq n$ and $1 \leq j \leq n$:

$$\alpha_{ij} = \begin{cases} 1 & \text{if } i \text{ precedes } j \text{ in } \Gamma_+, \\ 0 & \text{otherwise,} \end{cases}$$

$$\beta_{ij} = \begin{cases} 1 & \text{if } i \text{ precedes } j \text{ in } \Gamma_-, \\ 0 & \text{otherwise.} \end{cases}$$

The following proposition is straightforward to check and is illustrated in Fig. 4.4.

Proposition 4.1 *For any two departments i and j, the following hold:*

- *if $\alpha_{ij} = 1$ and $\beta_{ij} = 1$, then i is to the left of j;*
- *if $\alpha_{ij} = 0$ and $\beta_{ij} = 0$, then i is to the right of j;*
- *if $\alpha_{ij} = 0$ and $\beta_{ij} = 1$, then i is below j;*
- *if $\alpha_{ij} = 1$ and $\beta_{ij} = 0$, then i is above j.*

For example, consider the layout and sequence-pair in Fig. 4.3. Department 4 precedes department 5 in both Γ_+ and Γ_-, hence $(\alpha_{4,5}, \beta_{4,5}) = (1, 1)$, which correctly reflects the relationship between 4 and 5 according to Fig. 4.4, i.e., department 5 is to the right of department 4. By contrast, department 4 precedes department 2 only in Γ_+, so $(\alpha_{4,2}, \beta_{4,2}) = (1, 0)$, i.e., department 2 is below department 4 as per Fig. 4.4.

Fig. 4.4 Values of $(\alpha_{ij}, \beta_{ij})$ for a sequence-pair and possible locations for j with respect to i

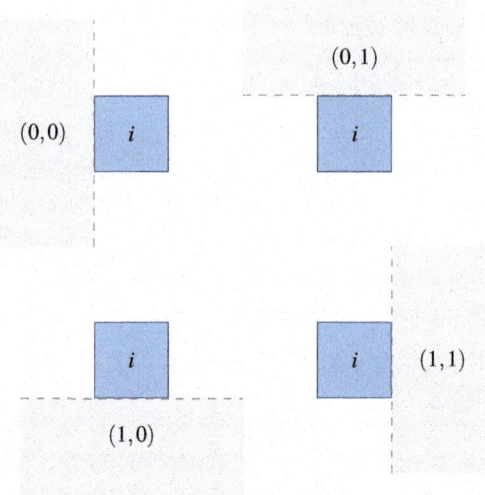

The sequence-pair-based MISOCO formulation for the UA-FLP is as follows:

minimize $\displaystyle\sum_{1\leq i<j\leq n} c_{ij}(d_{ij}^x + d_{ij}^y)$

s.t. $(4.13)-(4.20)$

$$\alpha_{ij} + \alpha_{ji} = 1, \quad 1 \leq i < j \leq n$$
$$\beta_{ij} + \beta_{ji} = 1, \quad 1 \leq i < j \leq n \tag{4.25}$$

$$\alpha_{ik} + \alpha_{kj} - \alpha_{ij} \leq 1, \quad 1 \leq i < j \leq n$$
$$\beta_{ik} + \beta_{kj} - \beta_{ij} \leq 1, \quad 1 \leq i < j \leq n \tag{4.26}$$

$$x_i + \tfrac{1}{2}w_i \leq x_j - \tfrac{1}{2}w_j + w_F(2 - \alpha_{ij} - \beta_{ij}), \quad 1 \leq i, j \leq n, \ i \neq j$$
$$y_i + \tfrac{1}{2}h_i \leq y_j - \tfrac{1}{2}h_j + h_F(1 + \alpha_{ij} - \beta_{ij}), \quad 1 \leq i, j \leq n, \ i \neq j \tag{4.27}$$

$$\alpha_{ij}, \beta_{ij} \in \{0, 1\}, \quad 1 \leq i, j \leq n. \tag{4.28}$$

Constraints (4.25) ensure that every department appears exactly once in each sequence, and constraints (4.26) are transitivity constraints that must hold for each of the two sequences. Together these constraints ensure that the binary variables represent valid sequences. Constraints (4.27) express the nonoverlapping requirement in terms of the binary variables (and hence of the sequence-pair).

4.4 Two-Stage Approaches

Two-stage approaches for the UA-FLP do not guarantee to find the global optimal solution. They are a compromise by which we sacrifice the ability to achieve guaranteed global optimality for the sake of being able to compute good solutions for large-scale instances. We lose the guarantee because the first stage finds a local (not necessarily global) minimum.

The first stage determines the relative positions of the departments in a desirable layout. In other words, the output of the first stage allows us to fix the values of the binary variables in the formulation in Sect. 4.2 or to fix the sequence-pairs in the formulation in Sect. 4.3.

The second stage uses the information from the first stage to determine the actual layout. The formulations in Sects. 4.2 and 4.3 become SOCOs after the binary variables are fixed, and such problems can be solved efficiently to global optimality by state-of-the-art commercial solvers.

The quality of a two-stage approach can be estimated by taking instances for which we know the optimal value (and perhaps one or more corresponding optimal layouts) and comparing this information to the best results obtained using the two-stage approach.

4.4.1 First Stage: Method Based on Nonlinear Optimization

In the next sections we describe two methods for the first stage. The first method relies on a nonlinear optimization problem, and the second uses a genetic algorithm.

The nonlinear optimization method handles the nonoverlapping requirement indirectly by penalizing overlap in the objective function. The idea is to use an *attractor–repeller paradigm* in which the objective function combines an attractor component and a repeller component:

- The attractor component is a function that seeks to make the distances between departments as small as possible by attracting all pairs of departments to each other. This component aims for a total distance of zero, i.e., a solution in which all the departments have their centres at the same point and thus fully overlap.
- The repeller component counteracts the effect of the attractor by seeking to enforce nonoverlap. It can take different forms, and the general idea is that the more the departments overlap, the higher the value of the repeller component.

Specifically, let

$$D_{ij} = (x_i - x_j)^2 + (y_i - y_j)^2$$

be the squared Euclidean distance between the centres of departments i and j. The attractor component is the sum of the connectivities c_{ij} multiplied by D_{ij}:

$$\sum_{ij} c_{ij} D_{ij}.$$

Possible choices for the repeller component are functions such as $f(t) = \frac{1}{t}$ or $f(t) = \log(t)$ for which $f(D_{ij}) \to +\infty$ when $D_{ij} \to 0$.

The nonlinear optimization problem representing this attractor–repeller paradigm is

$$\text{minimize} \quad \sum_{ij} \left(c_{ij} D_{ij} + K_{ij} \, f(D_{ij}) \right) \tag{4.29}$$

$$\begin{aligned}
\text{s.t.} \quad &\tfrac{1}{2}(w_i - w_F) \le x_i \le \tfrac{1}{2}(w_F - w_i), \quad 1 \le i < j \le n, \\
&\tfrac{1}{2}(h_i - h_F) \le y_i \le \tfrac{1}{2}(h_F - h_i), \quad 1 \le i < j \le n,
\end{aligned} \tag{4.30}$$

$$w_i \le \rho_i h_i \quad \text{and} \quad h_i \le \rho_i w_i, \quad 1 \le i \le n, \tag{4.31}$$

$$\begin{aligned}
&w_i^{\min} \le w_i \le w_i^{\max}, \quad 1 \le i < j \le n, \\
&h_i^{\min} \le h_i \le h_i^{\max}, \quad 1 \le i < j \le n,
\end{aligned} \tag{4.32}$$

$$w_i h_i \ge A_i, \quad 1 \le i \le n, \tag{4.33}$$

where the parameters K_{ij} aim to adequately balance the influence of the repeller component and that of the attractor component for each pair of departments.

A particularly effective implementation of the attractor–repeller paradigm uses the repeller function

$$f(D_{ij}) = \frac{\theta_{ij}^2}{D_{ij}} - 1,$$

where $\theta_{ij}^2 = \frac{1}{4}\left[(w_i + w_j)^2 + (h_i + h_j)^2\right]$ and $K = \alpha \sum_{1 \le i < j \le n} c_{ij}$, and $0 < \alpha \le 1$. Note that $D_{ij}/\theta_{ij} \approx 1$ indicates that some of the borders of the departments are close, regardless of whether or not the departments overlap (by a small amount). Using different choices of α (and hence of K), a variety of solutions for the first stage can be computed using a nonlinear optimization solver.

Once the nonlinear optimization problem has been solved, we want to use the information from the (local) optimal solution to extract information about how the departments should be located within the facility. The idea is to take the optimal positions of the centres of the departments according to the solution of the first stage as points on the plane and to determine the relative positions of the departments.

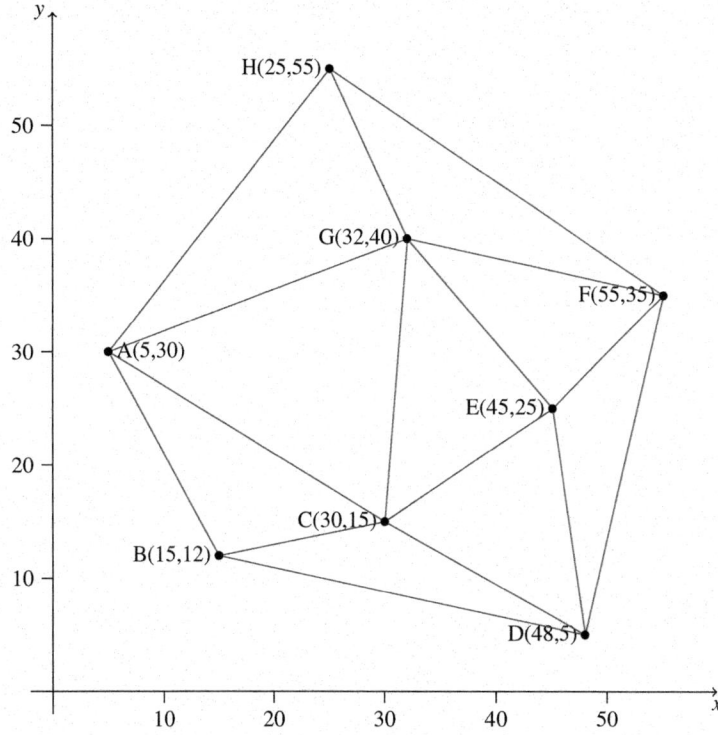

Fig. 4.5 Delaunay triangulation on a set of eight points

To achieve this, we use a triangulation of the plane. A triangulation is a division of a polygon into a set of triangles with the restriction that any two adjacent triangles share one side entirely. We are interested in triangulations that have all the vertices of the triangles at one of the points corresponding to the department centres. Furthermore, we use a specific triangulation called the Delaunay triangulation that has the property of maximizing the minimum angle of all the angles of the triangles in the triangulation. In practice, such a triangulation is unlikely to have thin triangles, and in our context, this results in a better space allocation to the departments. The Delaunay triangulation for a set of eight points is illustrated in Fig. 4.5.

Once we have computed the triangulation, we take the edges of the triangulation to represent the relative positions of the departments, and these positions are then enforced by appropriately fixing the binary variables $(\alpha_{ij}, \beta_{ij})$. In terms of the model in Sect. 4.2, this is equivalent, for each pair of departments i and j, to replacing the constraints (4.21) and (4.22) by a single linear constraint that achieves the relative positioning of these two departments in the model for the second stage, in accordance with the edges of the triangulation. Specifically, suppose that the centres of i and j are connected by an edge in the Delaunay triangulation. We have the following two requirements:

- If $|x_i - x_j| \geq |y_i - y_j|$ and $x_i \geq x_j$, then we enforce that i is to the right of j, so the second-stage model will include the linear constraint

$$x_j + \frac{1}{2}w_j \leq x_i - \frac{1}{2}w_i.$$

  However, if $x_i < x_j$, then we require that i is to the left of j, and we instead use the constraint

$$x_i + \frac{1}{2}w_i \leq x_j - \frac{1}{2}w_j.$$

- If $|y_i - y_j| \geq |x_i - x_j|$ and $y_i \geq y_j$, then we want to place i above j, so we use the constraint

$$y_j + \frac{1}{2}h_j \leq y_i - \frac{1}{2}h_i,$$

  whereas if $y_i < y_j$, then we locate i below j:

$$y_i + \frac{1}{2}h_i \leq y_j - \frac{1}{2}h_j.$$

4.4.2 First Stage: Method Based on Genetic Algorithm

An alternative approach to establish the relative positions between departments is to use a genetic algorithm (GA). GAs are a type of evolutionary algorithm and are based on the concepts of genetics and natural selection. They can be used to generate high-quality solutions for optimization problems that are beyond the ability of exact algorithms.

A GA works with a population of individuals and a set of rules to simulate the process of natural selection in the sense that individuals with higher fitness can generate more offspring than others can. Each individual is described by a chromosome, which is mathematically represented as a vector with elements from a suitable set. For example, for layout problems, a possible layout can be represented using sequence-pairs, and the corresponding chromosome for that layout would be a vector containing the sequences Γ_+ and Γ_- from its sequence-pair representation. A fitness score is given to each individual corresponding to its ability to compete. In the layout application, the fitness is the value of the objective function for that layout, i.e., the total connectivity cost between pairs of departments for the layout encoded in the chromosome.

Starting with a given initial generation of individuals, the chromosomes of each generation with better fitness scores are given more chance to reproduce than others, and this biases the composition of the subsequent generations. The algorithm mimics the evolution from one generation to another using the following operators:

- Selection: Select two chromosomes, giving preference to the individuals with higher fitness scores. Allow the selected chromosomes to pass their genes to subsequent generations.
- Crossover: Given two chromosomes, choose portions of each chromosome randomly and combine them so as to obtain a new valid chromosome.
- Mutation: Change portions of a chromosome randomly to increase the diversity of the population and seek chromosomes with better fitness.

In this way, the overall fitness of consecutive generations improves. Once the improvement in fitness between subsequent generations is sufficiently small, the algorithm has converged to a final generation, which is a set of solutions for the problem. In our application, this gives a population of layouts that includes the best layouts found by the algorithm.

For the UA-FLP, given the final population and the corresponding chromosomes with best fitness scores, if the chromosomes represent sequence-pairs, then we directly have the relative positions of the departments, and we can proceed to the second stage.

Sequence-pairs are not the only way to encode layouts in chromosomes. For example, we can also use the absolute position (x_i, y_i) and the shape parameter ρ_i of department i as the information in the chromosome using the vector $(x_1, y_1, \rho_1, x_2, y_2, \rho_2, \ldots, x_n, y_n, \rho_n)$. In this case, we need to recover the relative position of the departments. One way to do this is to use the Delaunay triangulation as described in Sect. 4.4.1. Alternatively, the variables $(\alpha_{ij}, \beta_{ij})$ can be assigned values according to the positioning of department j in one of the four sectors around department i, as depicted in Fig. 4.6.

Another possibility is to force the separation of departments in the direction in which they overlap less. To do this, we compute

$$\max\{x_i - x_j, x_j - x_i, y_i - y_j, y_j - y_i\}. \tag{4.34}$$

The expression giving the largest value establishes the direction in which we require the departments to be separated. The relative positions of the departments are obtained by making the variables $(\alpha_{ij}, \beta_{ij})$ equal to 0 or 1 according to Fig. 4.4.

Fig. 4.6 Relative position given by the chromosomes

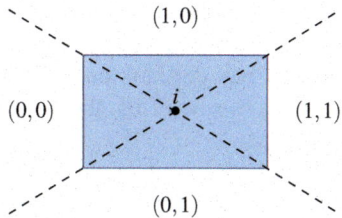

4.4.3 Second Stage

The second-stage model is a SOCO formulation of the UA-FLP in which the values of the variables $(\alpha_{ij}, \beta_{ij})$ have already been fixed based on the solution of the first stage. The second-stage model arises from the formulation in Sect. 4.2 and is as follows:

$$\text{minimize} \quad \sum_{1 \le i < j \le n} c_{ij}(d_{ij}^x + d_{ij}^y)$$

$$\begin{aligned}
\text{s.t.} \quad & d_{ij}^x \ge x_i - x_j, \quad d_{ij}^x \ge x_j - x_i, \quad 1 \le i < j \le n, \\
& d_{ij}^y \ge y_i - y_j, \quad d_{ij}^y \ge y_j - y_i, \quad 1 \le i < j \le n, \\
& h_i^{\min} \le h_i \le h_i^{\max}, \quad 1 \le i \le n, \\
& w_i^{\min} \le w_i \le w_i^{\max}, \quad 1 \le i \le n, \\
& w_i + h_i \ge \left\| \left(\frac{w_i - h_i}{2\sqrt{A_i}} \right) \right\|_2, \quad 1 \le i \le n \\
& w_i \le \rho_i h_i \quad \text{and} \quad h_i \le \rho_i w_i, \quad 1 \le i \le n, \\
& \frac{1}{2}(w_i - w_F) \le x_i \le \frac{1}{2}(w_F - w_i), \quad 1 \le i < j \le n, \\
& \frac{1}{2}(h_i - h_F) \le y_i \le \frac{1}{2}(h_F - h_i), \quad 1 \le i < j \le n, \\
& \text{linear nonoverlap constraints,}
\end{aligned}$$

where the linear nonoverlap constraints are obtained from (4.8) according to the solution of the first stage.

4.5 Flexible Bay Structure

A flexible bay structure is a version of the UA-FLP in which the departments are located in parallel bays with varying widths; this special case arises in the design of manufacturing facilities. We illustrate the concept of a flexible bay structure in Fig. 4.7. We observe that the column-wise bay structure is similar to the row structure in row FLPs. However, a fundamental difference is that the width of each flexible bay depends on the total area of the departments assigned to that bay, whereas in row FLPs, the heights of the rows (and of the departments) are equal and fixed.

The bays have straight aisles on both sides, and departments are not allowed to span multiple bays. This structure restricts the set of feasible solutions, but it has

Fig. 4.7 Facility layout with flexible bays: the width of each bay (column) depends on the assigned departments

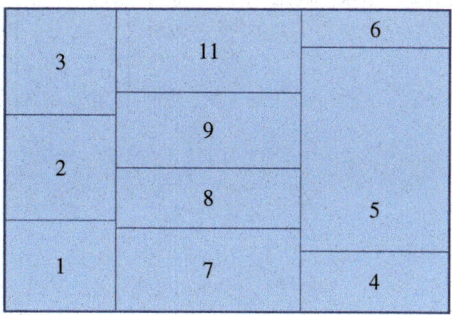

advantages in practice. In particular, the bay boundaries form the basis of an aisle structure that facilitates the transfer of the layout solution to an actual facility design.

We let the continuous variables x_i, y_i represent the location of department i, and we define h_{ik} to be the height of department i in bay k. Let K be the (given) set of bays. The binary variables are defined as follows:

$$z_{ik} = \begin{cases} 1, & \text{if department } i \text{ is assigned to bay } k \\ 0, & \text{otherwise;} \end{cases}$$

$$\alpha_{ij} = \begin{cases} 1, & \text{if department } i \text{ is above department } j \text{ in the same bay} \\ 0, & \text{otherwise;} \end{cases}$$

$$\delta_k = \begin{cases} 1, & \text{if bay } k \text{ is occupied} \\ 0, & \text{otherwise.} \end{cases}$$

A MILO model to optimize the flexible bay layout is as follows:

$$\text{minimize} \quad \sum_{1 \le i < j \le n} c_{ij} (d_{ij}^x + d_{ij}^y)$$

$$\text{s.t.} \quad d_{ij}^x \ge x_i - x_j, \quad d_{ij}^x \ge x_j - x_i, \quad 1 \le i < j \le n,$$

$$d_{ij}^y \ge y_i - y_j, \quad d_{ij}^y \ge y_j - y_i, \quad 1 \le i < j \le n,$$

$$\sum_{k \in K} z_{ik} = 1, \quad 1 \le i \le n, \tag{4.35}$$

$$w_k = \frac{1}{h_F} \sum_{i=1}^{n} z_{ik} A_i, \quad k \in K, \tag{4.36}$$

$$w_i^{\min} z_{ik} \le w_k \le w_i^{\max} + w_F(1 - z_{ik}), \quad 1 \le i \le n, \ k \in K \tag{4.37}$$

$$x_i \ge \sum_{j \le k} w_j - \tfrac{1}{2} w_k - (w_F - w_i^{\min})(1 - z_{ik}), \quad 1 \le i \le n, \ k \in K,$$

$$x_i \le \sum_{j \le k} w_j - \tfrac{1}{2} w_k + (w_F - w_i^{\min})(1 - z_{ik}), \quad 1 \le i \le n, \ k \in K, \tag{4.38}$$

$$\frac{h_{ik}}{A_i} - \frac{h_{jk}}{A_j} - \max\left\{\frac{h_i^{\max}}{A_i}, \frac{h_j^{\min}}{A_j}\right\}(2 - z_{ik} - z_{jk}) \leq 0, \quad 1 \leq i < j \leq n,$$

$$\frac{h_{ik}}{A_i} - \frac{h_{jk}}{A_j} + \max\left\{\frac{h_i^{\max}}{A_i}, \frac{h_j^{\min}}{A_j}\right\}(2 - z_{ik} - z_{jk}) \geq 0, \quad 1 \leq i < j \leq n,$$

$$(4.39)$$

$$\sum_{i=1}^{n} h_{ik} = h_F \delta_k, \quad k \in K, \tag{4.40}$$

$$h_i^{\min} z_{ik} \leq h_{ik} \leq h_i^{\max} z_{ik}, \quad 1 \leq i \leq n, \, k \in K, \tag{4.41}$$

$$\sum_{k \in K} h_{ik} = h_i, \quad 1 \leq i \leq n, \tag{4.42}$$

$$y_i - \frac{1}{2}h_i \geq y_j + \frac{1}{2}h_j - w_H(1 - r_{ij}), \quad 1 \leq i \neq j \leq n, \tag{4.43}$$

$$\alpha_{ij} + \alpha_{ji} \leq 1, \quad 1 \leq i < j \leq n, \tag{4.44}$$

$$\alpha_{ij} + \alpha_{ji} \geq z_{ik} + z_{jk} - 1, \quad 1 \leq i < j \leq n, \, k \in K, \tag{4.45}$$

$$\frac{1}{2}h_i \leq y_i \leq w_H - \frac{1}{2}h_i, \quad 1 \leq i \leq n. \tag{4.46}$$

Constraints (4.35) ensure that each department is assigned to a single bay. Constraints (4.36) calculate the width of each bay as the total area of the departments assigned to that bay divided by the facility height. This ensures that the bay width w_k is determined so that $w_k \times h_F$ has enough space to accommodate all the departments assigned to bay k. Note that under the assumption that $\sum_{i \in N} A_i \leq w_F h_F$, we have $\sum_{k \in K} w_k \leq w_F$. Constraints (4.37) impose bounds on the bay widths, based on the width bounds of the departments assigned to each bay.

Constraints (4.38) determine the horizontal locations of the department centres. In this model, the origin for the x coordinate is located in the middle of the bays. Therefore, if department i is assigned to bay k, x_i is calculated as $x_i = \sum_{j=1}^{k} w_j - \frac{1}{2}w_k$. This is consistent with constraints (4.38) when $z_{ik} = 1$.

Constraints (4.39) ensure that if departments i and j are in the same bay k, then their heights are determined according to the width of their assigned bay. Note that in this case both h_{ik}/A_i and h_{jk}/A_j must be equal to $1/w_k$. Moreover, we observe that constraints (4.39) are linear in the variables of the model.

Constraints (4.40) set the total heights of the departments in a bay equal to h_F if the bay is used and zero if the bay is empty. Constraints (4.41) are bounds on the department heights. They also enforce $h_{ik} = 0$ when department i is not located in bay k. Constraints (4.42) compute the height of each department. Constraints (4.43) prevent departments in the same bay from overlapping. Constraints (4.44)–(4.45) ensure that department i is either above or below department j. Finally, constraints (4.46) ensure that the departments are inside the facility.

4.6   References and Further Reading

The UA-FLP has received much attention since it was first stated in Armour and Buffa (1963). A MILO model for the problem was introduced by Meller et al (1999) and enhanced in Sherali et al (2003). In the latter work, the area constraint is replaced by a polyhedral outer approximation on Δ points of (4.4):

$$a_i w_i + 4 \left(w_i^{\min} + \frac{\lambda}{\Delta - 1} (w_i^{\max} - w_i^{\min}) \right)^2 h_i \geq 2 a_i \left(w_i^{\min} + \frac{\lambda}{\Delta - 1} (w_i^{\max} - w_i^{\min}) \right),$$

for $\lambda = 0, 1, \ldots, \Delta - 1$. This approximation was also used in Meller et al (2007) and Liu and Meller (2007). It is effective in practice but less efficient than the second-order cone relaxation in Sect. 4.1. Moreover, second-order cone constraints are supported by all state-of-the-art MILO solvers.

The sequence-pair representation was first used for circuit design in Murata et al (1995) and for the FLP in Meller et al (2007) and Liu and Meller (2007).

The attractor–repeller paradigm is from Anjos and Vannelli (2002). The strategy of measuring the centre-to-centre distance was presented in Jankovits et al (2011) and Anjos and Vieira (2016). The definition of four sectors to establish the relative position of the departments is due to Kulturel-Konak and Konak (2013). The Delaunay triangulation is a standard construction in computational geometry; more details can be found in Preparata and Shamos (2012). An accessible introduction to genetic algorithms can be found in the excellent book of Mitchell (1998). Using a two-stage approach closely following the ideas in this chapter, Anjos and Vieira (2016) computed layouts for instances with up to 100 departments in less than 15 min of computational time. Flexible bay structures were considered in Meller (1997), and a MILO formulation was proposed in Konak et al (2006).

References

Anjos MF, Vannelli A (2002) An attractor-repeller approach to floorplanning. Math Methods Oper Res 56(1):3–27

Anjos MF, Vieira MVC (2016) An improved two-stage optimization-based framework for unequal-areas facility layout. Optimization Letters 10(7):1379–1392

Armour GC, Buffa ES (1963) A heuristic algorithm and simulation approach to relative location of facilities. Management Science 9(2):294–309

Jankovits I, Luo C, Anjos MF, Vannelli A (2011) A convex optimisation framework for the unequal-areas facility layout problem. Eur J Oper Res 214(2):199–215

Konak A, Kulturel-Konak S, Norman BA, Smith AE (2006) A new mixed integer programming formulation for facility layout design using flexible bays. Oper Res Lett 34:660–672

Kulturel-Konak S, Konak A (2013) Linear programming based genetic algorithm for the unequal area facility layout problem. Int J Prod Res 51(14):4302–4324

Liu Q, Meller RD (2007) A sequence-pair representation and MIP-model-based heuristic for the facility layout problem with rectangular departments. IIE Transactions 39(4):377–394

Meller RD (1997) The multi-bay manufacturing facility layout problem. Int J Prod Res 35(5):1229–1237

Meller RD, Narayanan V, Vance PH (1999) Optimal facility layout design. Oper Res Lett 23:117–127

Meller RD, Chen W, Sherali HD (2007) Applying the sequence-pair representation to optimal facility layout designs. Oper Res Lett 35:651–659

Mitchell M (1998) An introduction to genetic algorithms. MIT Press

Murata H, Fujiyoshi K, Nakatake S, Kajitani Y (1995) Rectangle-packing-based module placement. In: Proceedings of IEEE international conference on computer aided design (ICCAD), pp 472–479

Preparata FP, Shamos MI (2012) Computational geometry: An introduction. Springer Science & Business Media

Sherali HD, Fraticelli BMP, Meller RD (2003) Enhanced model formulations for optimal facility layout. Operations Research 51(4):629–644

Chapter 5
Extensions and Related Problems

In this chapter, we introduce the quadratic assignment problem, a well-known special case of facility layout. We also briefly discuss the extensions of facility layout to re-layout, multi-floor layout, and dynamic versions of facility layout.

5.1 Quadratic Assignment Problem

The quadratic assignment problem (QAP) is a combinatorial optimization problem that dates back to 1957. It was developed as a mathematical model to assign economic activities to locations in an optimal way. Since then, it has been used in a wide variety of contexts, including facility layout.

To formulate facility layout as a QAP, we assume that we have n departments to assign to n locations and that *any department can be assigned to any location.* Because the locations are given and fixed, we also assume that the distance between every pair of locations is known. Let d'_{pq} denote the distance between locations p and q, for $1 \leq p, q \leq n$. Assuming that we are also given non-negative pairwise connectivities c_{ij} between departments i and j, then the cost of simultaneously assigning department i to location p and department j to location q is equal to $c_{ij} d'_{pq}$. Under these assumptions, the QAP seeks the permutation of the departments that minimizes the total weighted distance travelled, and analogously to formulation (2.6) for the SRFLP, it can be formulated as

$$\min_{\pi \in \Pi_n} \sum_{i=1}^{n} \sum_{j=1}^{n} c_{ij} d'_{\pi(i)\pi(j)}, \tag{5.1}$$

where Π_n is the set of all permutations π of $\{1, 2, \ldots, n\}$ and $\pi(i)$ is the location assigned to department i by permutation π.

© Springer Nature Switzerland AG 2021
M. F. Anjos, M. V. C. Vieira, *Facility Layout*, EURO Advanced Tutorials
on Operational Research, https://doi.org/10.1007/978-3-030-70990-7_5

The QAP can be expressed as a quadratic binary optimization problem. We define the binary variables

$$x_{ip} = \begin{cases} 1 \text{ if department } i \text{ is assigned to location } p, \\ 0 \text{ otherwise.} \end{cases}$$

Using these variables, the formulation of the QAP is as follows:

$$\text{minimize} \quad \sum_{i=1}^{n}\sum_{j=1}^{n}\sum_{p=1}^{n}\sum_{q=1}^{n} c_{ij}d'_{pq}x_{ip}x_{jq} \tag{5.2}$$

$$\text{s.t.} \quad \sum_{p=1}^{n} x_{ip} = 1, \quad 1 \le i \le n \tag{5.3}$$

$$\sum_{i=1}^{n} x_{ip} = 1, \quad 1 \le p \le n \tag{5.4}$$

$$x_{ip} \in \{0, 1\}, \quad 1 \le i, p \le n. \tag{5.5}$$

In the objective function, if department i is assigned to location p and department j is assigned to location q, then the product of c_{ij} and d'_{pq} is counted towards the total cost of the layout.

Constraints (5.3) ensure that each department is assigned to exactly one location, and constraints (5.4) ensure that precisely one department is assigned to each location. Constraints (5.3) and (5.4) together define an assignment of departments to locations. The assignment problem (5.2)–(5.5) is called quadratic because the objective function is quadratic in the variables x.

The QAP is well known for being computationally demanding. In general, it is challenging to solve a QAP with more than 30 departments to global optimality.

5.2   Re-Layout Problems

Re-layout refers to the process of making changes to an existing facility layout by relocating a certain number of departments to new locations, with possible changes to their area requirements, their aspect ratio requirements, or the bounds on their dimensions. For example, in the manufacturing context, the purpose of re-layout is to improve the workflow and, hence, the productivity of the facility, after changes to the product mix and/or the manufacturing processes themselves. Re-layout for a single floor has received the most attention, but re-layout applies to all types of layout problems.

One of the key features of re-layout is that it is constrained by some of the features of the existing layout, in particular the fact that certain departments (often corridors and elevators) cannot be moved. We assume that the available locations and their dimensions (and hence their areas) are known. We also assume that the area requirements of the departments to be relocated are known (these may be different from the original requirements). Given the information that relocated department i needs area A_i and location p has area B_p, we can add constraints of the form $A_i x_{ip} \leq B_p$ to any of the approaches in Chap. 4.

A further possible simplification is to assume that any of the departments to be relocated can be assigned to any of the available locations. In other words, we neglect all considerations about the area or shape of either departments or locations. It is straightforward to formulate this simplified version of re-layout as a QAP using the formulation (5.2)–(5.5) in Sect. 5.1, but re-layout (or facility layout itself) remains a difficult problem.

We have assumed that there is an equal number of departments and locations. If we have an instance with more locations than departments, then some locations will be left unassigned. This can be accommodated in the formulations in Chap. 4 and in the QAP model by changing constraint (5.4) to an inequality, so that at most one department is assigned to each location:

$$\sum_{i=1}^{n} x_{ip} \leq 1, \quad 1 \leq p \leq n.$$

5.3 Layout on Several Floors

The multi-floor FLP (MF-FLP) involves finding the optimal arrangement of departments in a facility with multiple floors. Practical applications include production facilities, hotels, office buildings, and hospitals. This problem is more complex than the UA-FLP because we must consider the interactions between departments on different floors. Furthermore, elevators and/or stairwells are required to transfer people and/or materials between the floors, and these need to be placed at coherent locations across floors.

5.3.1 MISOCO Formulation of the MF-FLP

We assume that the following parameters are given:

- the number of departments and their areas,
- the interconnection costs between pairs of departments,
- the number of floors,
- the dimensions and height of the floors, and
- the number and size of the elevators.

The number of floors and elevators is assumed to be fixed; if necessary, we could run the model for several different options. The floor dimensions are also assumed to be fixed, but they could easily be treated as decision variables.

For simplicity in the formulation, and without loss of generality, we assume that the elevators represent all the necessary systems for vertical movement between floors, including actual elevators, stairs, water pipes, and air ducts. For this reason, the dimensions of the elevators and/or restrictions on their possible locations are frequently given as part of the problem description.

We consider the problem of determining the optimal locations of the elevators, and the optimal locations and dimensions of the departments. The horizontal distance is the rectilinear distance (which is a reliable measure, as in the single-floor case), but the vertical distance must take into account the use of the elevators. This makes the formulation more complex.

Let δ be the ceiling height, p the number of floors, and e the number of elevators. Set $M = w_F + h_F + \delta p$, and define the following variables:

$z_{ik} = 1$ if department i is assigned to floor k, 0 otherwise;
$Z_{ij} = 1$ if departments i and j are allocated to the same floor, 0 otherwise;
X_{ij}, Y_{ij}: binary variables used to set up the nonoverlap constraints;
(x_i, y_i): coordinates of the centre of department i;
d_{ij}^v: vertical distance between departments i and j;
d_{ij}^h: horizontal distance between departments i and j;

where the indices $1, \ldots, n$ correspond to the departments, and the indices $n + 1, \ldots, n + e$ correspond to the elevators.

We can formulate the MF-FLP as follows:

$$\text{minimize} \quad \sum_{1 \le i < j \le n} c_{ij}(d_{ij}^h + d_{ij}^v) \tag{5.6}$$

$$\text{s.t.} \quad \sum_{k=1}^{p} z_{ik} = 1, \quad 1 \le i \le n \tag{5.7}$$

$$Z_{ij} \ge z_{ik} + z_{jk} - 1, \quad 1 \le i < j \le n, \ k = 1, \ldots, p \tag{5.8}$$

$$Z_{ij} \le 1 - z_{ik} + z_{jk}, \quad 1 \le i < j \le n, \ k = 1, \ldots, p \tag{5.9}$$

$$Z_{ij} \le 1 + z_{ik} - z_{jk}, \quad 1 \le i < j \le n, \ k = 1, \ldots, p \tag{5.10}$$

$$d_{ij}^v = \delta \left| \sum_{k=1}^{p} k(z_{ik} - z_{jk}) \right|, \quad 1 \le i < j \le n \tag{5.11}$$

$$d_{ij}^h \ge |x_i - x_j| + |y_i - y_j|, \quad 1 \le i < j \le n \tag{5.12}$$

$$d_{ij}^h \ge |x_i - x_\ell| + |y_i - y_\ell| + |x_j - x_\ell| + |y_j - y_\ell| - 2(w_F + h_F) Z_{ij}, \tag{5.13}$$

$$1 \leq i < j \leq n, \ n+1 \leq \ell \leq n+e$$

$$x_i + \frac{1}{2} w_i \leq \frac{1}{2} w_F, \ x_i - \frac{1}{2} w_i \geq -\frac{1}{2} w_F, \quad 1 \leq i \leq n+e \tag{5.14}$$

$$y_i + \frac{1}{2} h_i \leq \frac{1}{2} h_F, \ y_i - \frac{1}{2} h_i \geq -\frac{1}{2} h_F, \quad 1 \leq i \leq n+e \tag{5.15}$$

$$w_i h_i \geq A_i, \quad 1 \leq i \leq n \tag{5.16}$$

$$w_i - \beta h_i \leq 0, \ h_i - \beta w_i \leq 0, \quad 1 \leq i \leq n \tag{5.17}$$

$$x_i - x_j \geq \frac{1}{2}(w_i + w_j) - w_F(1 - Z_{ij} + X_{ij} + Y_{ij}), \quad 1 \leq i < j \leq n+e \tag{5.18}$$

$$x_j - x_i \geq \frac{1}{2}(w_i + w_j) - w_F(2 - Z_{ij} - X_{ij} + Y_{ij}), \quad 1 \leq i < j \leq n+e \tag{5.19}$$

$$y_i - y_j \geq \frac{1}{2}(h_i + h_j) - h_F(2 - Z_{ij} + X_{ij} - Y_{ij}), \quad 1 \leq i < j \leq n+e \tag{5.20}$$

$$y_j - y_i \geq \frac{1}{2}(h_i + h_j) - h_F(3 - Z_{ij} - X_{ij} - Y_{ij}), \quad 1 \leq i < j \leq n+e \tag{5.21}$$

$$z_{ik} = 1, \quad n+1 \leq i \leq n+e, \ 1 \leq k \leq p \tag{5.22}$$

$$Z_{ij} = 1, \quad n+1 \leq i < j \leq n+e \tag{5.23}$$

$$X_{ij}, Y_{ij}, Z_{ij}, z_{ik} \in \{0, 1\}, \quad 1 \leq i < j \leq n+e, \ 1 \leq k \leq p \tag{5.24}$$

$$h_i, w_i \geq 0, \quad 1 \leq i \leq n. \tag{5.25}$$

Note that as stated above, this formulation is not a MISOCO problem because of constraints (5.11), (5.12), (5.13), and (5.16). However, the first three of these can be linearized as explained in Sect. 2.3.1, and the fourth set is equivalent to the SOC constraints (4.11). After these adjustments, the result is a MISOCO formulation of MF-FLP.

Constraints (5.7) allocate each department to exactly one floor. Constraints (5.8), (5.9), and (5.10) set $Z_{ij} = 1$ if i and j are on the same floor, and 0 otherwise. Constraints (5.11) compute the vertical distance between each pair of departments. Note that

$$\left| \sum_{k=1}^{p} k(z_{ik} - z_{jk}) \right|$$

is equal to the number of floors separating departments i and j.

Constraints (5.12) and (5.13) compute the horizontal distance between each pair of departments. If departments i and j are on different floors, then the distance

d_{ij}^h includes the use of an elevator. Otherwise, $Z_{ij} = 1$ so this constraint becomes inactive and d_{ij}^h is found by constraint (5.12). Constraints (5.14), (5.15), (5.16), and (5.17) were discussed in Sect. 4.1.

Constraints (5.18)–(5.21) prevent the overlapping of departments and elevators on the same floor. To see how these constraints operate, suppose first that $Z_{ij} = 0$. This means that departments i and j are on different floors, and therefore we do not want to enforce a nonoverlap constraint between them. Observe that because $Z_{ij} = 0$, for the four possible assignments of binary values to X_{ij} and Y_{ij}, we have

$$(1 - Z_{ij} + X_{ij} + Y_{ij}) = 1 + X_{ij} + Y_{ij} > 0,$$
$$(2 - Z_{ij} - X_{ij} + Y_{ij}) = 2 - X_{ij} + Y_{ij} > 0,$$
$$(2 - Z_{ij} + X_{ij} - Y_{ij}) = 2 + X_{ij} - Y_{ij} > 0,$$
$$(3 - Z_{ij} - X_{ij} - Y_{ij}) = 3 - X_{ij} - Y_{ij} > 0.$$

Therefore, the right-hand side of constraints (5.18)–(5.21) will always be inactive.

Alternatively, if $Z_{ij} = 1$, then departments i and j are on the same floor, and therefore we must enforce one of the four nonoverlap constraints. Because $Z_{ij} = 1$, for any assignment of binary values to X_{ij} and Y_{ij}, we have

$(1 - Z_{ij} + X_{ij} + Y_{ij}) = X_{ij} + Y_{ij}$, which will equal 0 only if $X_{ij} = 0$ and $Y_{ij} = 0$;

$(2 - Z_{ij} - X_{ij} + Y_{ij}) = 1 - X_{ij} + Y_{ij}$, which will equal 0 only if $X_{ij} = 1$ and $Y_{ij} = 0$;

$(2 - Z_{ij} + X_{ij} - Y_{ij}) = 1 + X_{ij} - Y_{ij}$, which will equal 0 only if $X_{ij} = 0$ and $Y_{ij} = 1$;

$(3 - Z_{ij} - X_{ij} - Y_{ij}) = 2 - X_{ij} - Y_{ij}$, which will equal 0 only if $X_{ij} = 1$ and $Y_{ij} = 1$.

Therefore, depending on the specific values of X_{ij} and Y_{ij}, precisely one of the right-hand sides of constraints (5.18)–(5.21) will equal 0, which makes that constraint active and ensures that the two departments do not overlap.

Finally, constraints (5.22) and (5.23) ensure that each elevator covers all the floors and every pair of elevators shares the same floor.

5.3.2 Two-Stage Approach for the MF-FLP

Another way to tackle the MF-FLP is to first allocate the departments to floors while minimizing the vertical interaction costs, and then optimize the layout of the floors independently as multiple instances of single-floor layout. This is essentially a two-stage approach, akin to those presented in Sect. 4.4 for single-floor layout.

The assignment of departments to floors can be done using a *quadratic semi-assignment problem (QSAP)*. This is a version of the QAP in which each department must be assigned to exactly one floor, but more than one department can be assigned to each floor. The QSAP formulation is as follows:

$$\text{minimize} \quad \sum_{i=1}^{n} \sum_{j=i+1}^{n} \sum_{k=1}^{p} \sum_{l=1}^{p} c_{ij} D_{kl} z_{ik} z_{jl} \tag{5.26}$$

$$\text{s.t.} \quad \sum_{k=1}^{p} z_{ik} = 1, \quad 1 \le i \le n \tag{5.27}$$

$$\sum_{i=1}^{n} A_i z_{ik} \le w_F \, h_F, \quad 1 \le k \le p \tag{5.28}$$

$$z_{ik} \in \{0, 1\}, \quad 1 \le i \le n, \ 1 \le k \le p. \tag{5.29}$$

The variable z_{ik} equals 1 if department i is assigned to floor k, and 0 otherwise. Here D_{kl} is not a variable, but a parameter prespecified via $D_{kl} = \delta |k - l|$. The distance between departments i and j is given by $D_{kl} z_{ik} z_{jl}$. This quantity is set to D_{kl} if i and j are assigned to floors k and l, respectively, and zero otherwise. Constraints (5.27) assign each department to exactly one floor. Constraints (5.28) ensure that the departments assigned to each floor fit into that floor.

One limitation of this formulation is that the distance is assumed to be measured vertically through floors, which may be problematic in practical applications. An alternative is a MILO formulation. Let the variables d_{ij}^v represent the vertical distance between each pair i, j of departments, and consider the following MILO model:

$$\text{minimize} \quad \sum_{1 \le i < j \le n} c_{ij} d_{ij}^v \tag{5.30}$$

$$\text{s.t.} \quad \sum_{k=1}^{p} z_{ik} = 1, \ 1 \le i \le n \tag{5.31}$$

$$d_{ij}^v \ge \delta \sum_{k=1}^{p} k(z_{ik} - z_{jk}), \ 1 \le i < j \le n \tag{5.32}$$

$$d_{ij}^v \ge \delta \sum_{k=1}^{p} k(z_{jk} - z_{ik}), \ 1 \le i < j \le n \tag{5.33}$$

$$\sum_{i=1}^{n} A_i z_{ik} \le w_F \, h_F, \ 1 \le k \le p \tag{5.34}$$

$$z_{ik} \in \{0, 1\}, \ 1 \le i \le n, \ 1 \le k \le p. \tag{5.35}$$

The key here is that constraints (5.32)–(5.33) compute the vertical distance d_{ij}^v, where δ is the floor height. We recall that

$$\left| \sum_{k=1}^{p} k(z_{ik} - z_{jk}) \right|$$

is equal to the number of floors separating i and j.

In the second stage, the departments have been assigned to floors, and hence each floor becomes an instance of the UA-FLP with additional constraints to ensure coherence in the location of the elevators.

5.3.3 Multi-floor Layout with Evacuation Requirements

Facility layout problems can also be modelled as generalized QAPs. The *generalized quadratic assignment problem (GQAP)* is a variant of the QAP in which every location has an associated capacity, and any number of departments can be assigned to each location provided the capacity of that location is respected. We introduce this modelling approach by considering a specific version of the MF-FLP.

Suppose that we are given n departments to be allocated to p rectangular floors. The goal is, as usual, to minimize the total pairwise flow costs between departments subject to the following requirements:

1. No department can be split between different floors.
2. The available space on each floor cannot be exceeded.
3. The evacuation capacity of every floor must be respected to facilitate emergency evacuation.

The parameters specifying an instance of this MF-FLP with evacuation requirements are as follows:

- s is the number of stairwells;
- a_i is the area needed by department i;
- A_k is the available area on floor k;
- d_{kl} is the distance between floor k and floor l;
- λ_i is the average arrival rate of persons from department i during an evacuation;
- μ_d is the evacuation rate capacity of stairwell d.

We use the following binary variables:

- $x_{ik} = 1$ if department i is assigned to floor k, and 0 otherwise.
- $y_{id} = 1$ if department i is assigned to stairwell d, and 0 otherwise.

The GQAP model for the MF-FLP with evacuation requirements is

$$\text{minimize} \quad \sum_{ij} \sum_{kl} c_{ij} d_{kl} x_{ik} x_{jl} \tag{5.36}$$

$$\text{s.t.} \quad \sum_{k=1}^{p} x_{ik} = 1, \quad 1 \leq i \leq n, \tag{5.37}$$

$$\sum_{i=1}^{n} a_i x_{ik} \leq A_k, \quad 1 \leq k \leq p, \tag{5.38}$$

$$\sum_{d=1}^{s} y_{id} = 1, \quad 1 \leq i \leq n, \tag{5.39}$$

$$\sum_{i=1}^{n} \lambda_i y_{id} \leq \mu_d, \quad 1 \leq d \leq s, \tag{5.40}$$

$$x_{ik}, y_{id} \in \{0, 1\}, \quad 1 \leq i \leq n, \ 1 \leq k \leq p, \ 1 \leq d \leq s. \tag{5.41}$$

The objective function (5.36) is the same as (5.2). Constraints (5.37) assign each department to exactly one floor. Constraints (5.38) ensure that for each floor the available area is not exceeded. Constraints (5.39) assign each department to exactly one stairwell. Constraints (5.40) control the evacuation capacity of each stairwell by preventing the assignment of too many departments to that stairwell.

It may be of interest to simultaneously minimize the total distance travelled by evacuating residents

$$\sum_{i=1}^{n} \sum_{k=1}^{p} \sum_{d=1}^{s} \lambda_i e_{jd} x_{ik} y_{id}, \tag{5.42}$$

where e_{jd} is the distance between floor j and the exit through stairwell d. This evacuation distance may be added to the objective function (5.36) with an appropriate weight to balance its importance in relation to the minimization of the total pairwise flow costs between departments.

5.4 Dynamic Facility Layout

A dynamic approach to facility layout is useful in contexts where the dynamic nature of the practical application will require future changes to the layout. Some of the significant factors that may change over the planning horizon are as follows:

- changes in the design of products;
- addition or deletion of products;

- replacement of production equipment;
- shorter product life cycles;
- changes in the production quantities and associated production schedules.

The *dynamic facility layout problem (DFLP)* considers the design of layouts over several time periods, where demand volumes and product mix vary from one period to another. In other words, the potential for future changes may motivate appropriate planning, taking into account the expected changes in the costs and the additional rearrangement costs.

The general idea of the formulations we present below is to extend previous models from the single-period to the multi-period case. In principle, every mathematical optimization model presented in this book can be similarly extended to obtain its dynamic version.

5.4.1 QAP Formulation

We start by extending the QAP model (5.2–5.5). All the data considered previously is now assumed to be given for each period of a planning horizon of T periods. In addition, we must account for the rearrangement costs, so we let C_{ilk}^t denote the cost of changing department i from location l to location k in time period t. The binary variables are

$$x_{ik}^t = \begin{cases} 1, & \text{if department } i \text{ is assigned to location } k \text{ at period } t, \\ 0, & \text{otherwise.} \end{cases}$$

The resulting formulation is

$$\text{minimize} \quad \sum_{t=1}^{T} \sum_{i,j,l} c_{ij}^t d_{kl}^t x_{ik}^t x_{jl}^t + \sum_{t=2}^{T} \sum_{i,j,l} C_{ilk}^t x_{il}^{(t-1)} x_{ik}^t \tag{5.43}$$

$$\text{s.t.} \quad \sum_{k=1}^{n} x_{ik}^t = 1, \quad 1 \le i \le n,\ 1 \le t \le T, \tag{5.44}$$

$$\sum_{i=1}^{n} x_{ik}^t = 1, \quad 1 \le k \le n,\ 1 \le t \le T, \tag{5.45}$$

$$x_{ik}^t \in \{0,1\}, \quad 1 \le i,k \le n,\ 1 \le t \le T. \tag{5.46}$$

The quantity $d_{kl}^t x_{ik}^t x_{jl}^t$ is the distance between i and j if they are assigned to locations k and l, respectively. Similarly, the term $C_{ilk}^t x_{il}^{(t-1)} x_{ik}^t$ is the rearrangement cost of department i if it moves from location l to location k at the beginning of period t.

5.4.2 MISOCO Formulation

In this section we extend the MISOCO formulation for the UA-FLP presented in Sect. 4.2 to its dynamic version. We consider the same variables as before but duplicated for each period, and we add to the objective function terms for the rearrangement costs. We also have to control the position and shape changes of the departments.

We assume that we are given the following data:

- c_{ij}^t is the unit flow cost from department i to department j in period t;
- C_i^t is the cost of changing department i at the beginning of period t;
- $h_i^{\min}$ and $h_i^{\max}$ are lower and upper bounds on the height of department i;
- $w_i^{\min}$ and $w_i^{\max}$ are lower and upper bounds on the width of department i;
- w_F, h_F are the width and height of the facility;
- A_i is the required area for department i;
- ρ_i is the upper bound on the aspect ratio of department i.

The variables are as follows:

- (x_i^t, y_i^t) are the coordinates of the centre of department i in period t;
- $d^{xt}_{ij} + d^{yt}_{ij}$ gives the rectilinear distance between departments i and j in period t;
- w_i^t, h_i^t are the width and height of department i in period t;
- $\alpha_{ij}^t, \beta_{ij}^t$ determine the relative positions between departments i and j in period t, with the same interpretation as in Sect. 4.2;
- r_i^t is equal to 1 if department i is changed at the beginning of period t, and 0 otherwise.

The MISOCO formulation of the dynamic UA-FLP is as follows:

$$\text{minimize} \quad \sum_{t=1}^{T}\sum_{1\le i<j\le n} c_{ij}^t(d^{xt}_{ij}+d^{yt}_{ij}) + \sum_{t=2}^{T}\sum_{1\le i<j\le n} C_i^t r_i^t \tag{5.47}$$

$$d^{xt}_{ij} \ge x_i^t - x_j^t, \quad d^{xt}_{ij} \ge x_j^t - x_i^t, \quad 1 \le i < j \le n,\ 1 \le t \le T, \tag{5.48}$$

$$d^{yt}_{ij} \ge y_i^t - y_j^t, \quad d^{yt}_{ij} \ge y_j^t - y_i^t, \quad 1 \le i < j \le n,\ 1 \le t \le T, \tag{5.49}$$

$$h_i^{\min} \le h_i^t \le h_i^{\max}, \quad 1 \le i \le n,\ 1 \le t \le T, \tag{5.50}$$

$$w_i^{\min} \le w_i^t \le w_i^{\max}, \quad 1 \le i \le n,\ 1 \le t \le T, \tag{5.51}$$

$$w_i^t h_i^t \ge A_i, \quad 1 \le i \le n,\ 1 \le t \le T, \tag{5.52}$$

$$w_i^t \le \rho_i h_i^t \quad \text{and} \quad h_i^t \le \rho_i w_i^t, \quad 1 \le i \le n,\ 1 \le t \le T, \tag{5.53}$$

$$\frac{1}{2}(w_i^t - w_F) \le x_i^t \le \frac{1}{2}(w_F - w_i^t), \quad 1 \le i \le n,\ 1 \le t \le T, \tag{5.54}$$

$$\frac{1}{2}(h_i^t - h_F) \le y_i^t \le \frac{1}{2}(h_F - h_i^t), \quad 1 \le i \le n, \ 1 \le t \le T, \tag{5.55}$$

$$x_i^t + \frac{1}{2}w_i^t \le x_j^t - \frac{1}{2}w_j^t + w_F(1 - \alpha_{ij}^t), \quad 1 \le i \ne j \le n, \ 1 \le t \le T, \tag{5.56}$$

$$y_i^t + \frac{1}{2}h_i^t \le y_j^t - \frac{1}{2}h_j^t + h_F(1 - \beta_{ij}^t), \quad 1 \le i \ne j \le n, \ 1 \le t \le T, \tag{5.57}$$

$$\alpha_{ij}^t + \alpha_{ji}^t + \beta_{ij}^t + \beta_{ji}^t = 1, \quad 1 \le i < j \le n, \ 1 \le t \le T, \tag{5.58}$$

$$x_i^t - x_i^{(t-1)} \le w_F r_i^t, \quad -x_i^t + x_i^{(t-1)} \le w_F r_i^t, \quad 1 \le i \le n, \ 2 \le t \le T, \tag{5.59}$$

$$y_i^t - y_i^{(t-1)} \le h_F r_i^t, \quad -y_i^t + y_i^{(t-1)} \le h_F r_i^t, \quad 1 \le i \le n, \ 2 \le t \le T, \tag{5.60}$$

$$w_i^t - w_i^{(t-1)} \le w_F r_i^t, \quad -w_i^t + w_i^{(t-1)} \le w_F r_i^t, \quad 1 \le i \le n, \ 2 \le t \le T, \tag{5.61}$$

$$h_i^t - h_i^{(t-1)} \le h_F r_i^t, \quad -h_i^t + h_i^{(t-1)} \le h_F r_i^t, \quad 1 \le i \le n, \ 2 \le t \le T, \tag{5.62}$$

$$\alpha_{ij}^t, \beta_{ij}^t, r_i^t \in \{0, 1\}, \quad 1 \le i, j \le n, \ 1 \le t \le T. \tag{5.63}$$

For each period, almost all the constraints have the same interpretation as for the UA-FLP in Sect. 4.2. We recall that the constraints (5.52) can be expressed using a SOC of the form (4.11). The additional constraints specific to the dynamic version are (5.59)–(5.62). They work in combination with the variables r_i^t to control if there is some position or shape change for department i from one period to the next. If $r_i^t = 1$, then the cost of changing department i at the beginning of period t is included in the objective function; the coordinates of the centre of i in period t can be different from those in period $t - 1$; and the shape of i can change as well. If $r_i^t = 0$, then these quantities must remain unchanged from period $t - 1$ to period t.

5.5 References and Further Reading

The QAP as defined in (5.1) is known as the Koopmans–Beckmann QAP. The first statement of the QAP in the literature is found in Koopmans and Beckmann (1957), where the formulation (5.2)–(5.5) was also proposed. Çela (2013) is an excellent reference on the many aspects of the QAP, which remains a particularly challenging problem in discrete optimization.

Constraints (5.8)–(5.10) and (5.18)–(5.21) were taken from Patsiatzis and Papageorgiou (2002). Meller and Bozer (1997) present a QAP model to assign departments to floors. Bernardi and Anjos (2013) present a MILO formulation for the assignment of departments to floors as the first stage of a two-stage approach for MF-FLP.

The evacuation application of the MF-FLP was formulated as a GQAP in Hahn et al (2010), where it is called the multi-story assignment problem. We chose to use the term multi-floor to be consistent with the terminology of this book.

For more on dynamic FLP, we suggest Lahmar and Benjaafar (2005) and Zhu et al (2018). The factors affecting the connectivity costs (or material flows) are discussed in Shore and Tompkins (1980). Many authors have considered the QAP model for DFLP, for instance Ulutas and Islier (2015). A MILO formulation can be found in McKendall and Hakobyan (2010): the departments have fixed dimensions but are allowed to rotate.

References

Bernardi S, Anjos MF (2013) A two-stage mathematical-programming method for the multi-floor facility layout problem. J Oper Res Soc 64(3):352–364

Çela E (2013) The quadratic assignment problem: Theory and algorithms, vol 1. Springer Science & Business Media

Hahn P, MacGregor Smith J, Zhu YR (2010) The multi-story space assignment problem. Ann Oper Res 179(1):77–103

Koopmans TC, Beckmann M (1957) Assignment problems and the location of economic activities. Econometrica 25(1):53–76

Lahmar M, Benjaafar S (2005) Design of distributed layouts. IIE Transactions 37(4):303–318

McKendall AR, Hakobyan A (2010) Heuristics for the dynamic facility layout problem with unequal-area departments. Eur J Oper Res 201(1):171–182

Meller RD, Bozer YA (1997) Alternative approaches to solve the multi-floor facility layout problem. J Manuf Syst 16(3):192–203

Patsiatzis DI, Papageorgiou LG (2002) Optimal multi-floor process plant layout. Comput Chem Eng 26(4):575–583

Shore RH, Tompkins JA (1980) Flexible facilities design. AIIE Transactions 12(2):200–205

Ulutas B, Islier AA (2015) Dynamic facility layout problem in footwear industry. J Manuf Syst 36:55–61

Zhu T, Balakrishnan J, Cheng CH (2018) Recent advances in dynamic facility layout research. INFOR: Inf Syst Oper Res 56(4):428–456

Chapter 6
Engineering Applications of Facility Layout

This chapter gathers a variety of examples to illustrate the range of applications of facility layout. The main areas represented are manufacturing and hospital layout. This selection of examples is intended to give insight into the data that might need to be gathered and to illustrate that real data is not always available in a clear format. Some processing may need to be carried out before the data is ready for use with the mathematical optimization model.

We include applications for the various types of layout problems presented in the previous chapters. Specifically, we present applications for row layout, single-floor layout, multi-floor layout, dynamic layout, and the QAP. In some cases the nature of the layout problem is specified, and in others we leave it to the reader to determine how the structure of the problem and the data provided may be used to develop a suitable model.

The sources of these applications are given in Sect. 6.10, and we refer the interested reader to the references for more information about the examples.

6.1 Metallurgical Application

This first simple example is interesting because it comes from a master's thesis carried out in the context of a company more than 55 years ago, when computing power was limited to an extent that is now difficult to imagine. It illustrates the calculations and simplifications that had to be made to solve layout instances at that time. Despite the simplification, these instances were still useful in practice.

A metallurgical manufacturer needs to build a new plant in which all the departments (listed in Table 6.1) must be of equal size. The volume of parts to be handled, the capacity of the handling vehicles, and the sequence of departments that will handle each part are described in Table 6.2. For the distances between locations, we assume a 3×3 floor where each cell is a unit square.

© Springer Nature Switzerland AG 2021
M. F. Anjos, M. V. C. Vieira, *Facility Layout*, EURO Advanced Tutorials
on Operational Research, https://doi.org/10.1007/978-3-030-70990-7_6

Table 6.1 Departments for
the metallurgical application

	Department
(a)	Receiving
(b)	Snagging and Inspection
(c)	Milling
(d)	Automatic Screwing
(e)	Welding
(f)	Grinding
(g)	Plating
(h)	Painting
(i)	Packing and Warehouse

Table 6.2 Volume, handling capacity, and department sequence for each part

Part	Volume (pieces/year)	Handling capacity (pieces/load)	Sequence
1	5,000	20	(a) (b) (c) (h) (i)
2	12,000	200	(a) (c) (e) (g) (i)
3	600	30	(a) (b) (c) (f) (g) (i)
4	2,000	500	(a) (d) (e) (c) (i)
5	5,000	100	(a) (b) (h) (c) (f) (i)
6	9,000	50	(a) (d) (i)
7	20,000	1,000	(a) (d) (g) (i)
8	2,000	100	(a) (h) (f) (i)
9	1,000	250	(a) (b) (d) (g) (i)

The data must be manipulated to obtain the flows. For example, for part 1 the flow between departments (a) and (b) is $5000/20 = 250$. We perform the same calculation for all parts that are moved from (a) to (b) and sum the results to obtain the total flow between (a) and (b). This must be repeated for every pair of departments.

6.2   Flow-Line Layout in Manufacturing

In this example we consider 18 parts to be processed in 12 machines. The sequence of machines to be followed for each part and the production volume needed for each part are given in Table 6.3. The machines are to be arranged in a single row (see Chap. 2) so as to optimize the distance travelled by the parts in going from one machine to another.

Table 6.3 Production volume and sequence of machines for each part

Part	Production volume	Machine sequence
P01	100	M1-M4-M2-M6
P02	120	M3-M5-M12-M10
P03	50	M2-M4-M12-M6
P04	45	M5-M8-M10
P05	60	M3-M5-M12-M6
P06	80	M4-M2-M4-M6
P07	90	M1-M5-M9
P08	120	M3-M7-M10-M4-M8
P09	140	M1-M4-M6
P10	180	M3-M12-M8-M10
P11	80	M2-M6-M2-M4-M6
P12	60	M11-M9-M10-M8
P13	70	M1-M4-M5-M7
P14	150	M2-M4-M6-M2-M6
P15	120	M3-M7-M9-M10
P16	120	M3-M10-M12-M9-M12
P17	100	M5-M10-M8-M9-M12
P18	90	M2-M8-M9-M10

6.3 Application in the Defence Industry

This section presents one application among many in the area of defence: the production of wheel arm brackets for defence vehicles. There are two components per bracket and a total of eight machines; each machine carries out a specific operation for the manufacture of a bracket component. Table 6.4 lists the machines and processing times that each component requires, and Table 6.5 gives the specifications of the machines. This problem is an example of the UA-FLP (see Chap. 4). For each machine, the clearances required for safe operation are included in the total length and width.

Table 6.4 Machine types and component processing times

Machine number	Operation	Operation time (min)	
		Component 1	Component 2
1	Marking	60	45
2	Milling	60	60
3	VTL A	300	–
4	VTL B	90	100
5	HMC 1	150	120
6	HMC 2	135	–
7	Drilling	60	60
8	Benchwork	30	20

Table 6.5 Specifications of machines for production of wheel arm brackets

Machine number	Length	Width	Clearance length	Clearance width	Total length	Total width	Area
1	20	15	15	10	35	25	875
2	20	15	15	10	35	25	875
3	25	20	15	10	40	30	1200
4	30	25	15	10	45	35	1575
5	30	25	15	10	45	35	1575
6	10	10	15	10	25	20	500
7	25	20	15	10	40	30	1200
8	10	10	15	10	25	20	500

Observe that Component 1 is processed on all eight machines, while Component 2 is processed on six of the eight. However, the order in which these operations are to be carried out is not specified. By fixing different orders for the operations, different instances can be obtained, and in general their optimal layouts will differ. Note also that the dimensions of the facility are not provided. Different instances can be obtained by specifying these dimensions, or alternatively by leaving them to be determined by the optimization.

6.4 Shoe Manufacturing

In this section, we present an application from shoe manufacturing. Table 6.6 describes the processes in shoe production. Each process is subdivided into several steps, and each step is carried out at a workstation. The detailed production flow involves 2 warehouses and 52 workstations. The flows between workstations are given in Table 6.7. In this table, an entry of the form (23, 24) 26200 is interpreted as

Table 6.6 Process names and descriptions

Process number	Name	Definition
1	Clicking	The skins, hides, or man-made materials are cut into the shoe upper sections
2	Closing	The upper sections are stitched together to create the upper
3	Bottom-stock	The bottom components (such as soles and insoles) are cut and prepared
4	Lasting	The assembled uppers are gathered onto the last, and the upper is attached to the insole
5	Making	The heels and soles are attached to the lasted shoe
6	Finishing	Various operations such as bottom securing and edge trimming improve the durability and appearance of the shoe

Table 6.7 Workflows for the shoe factory

Flows (direction and amount)				
(wrm,1) 4350	(wrm,2) 1850	(wrm,14) 30200	(wrm,15) 34500	(wrm,44) 4900
(1,3) 4350	(2,4) 1850	(3,5) 4500	(4,5) 4000	(5,6) 8500
(6,7) 4250	(6,8) 4250	(7,9) 8500	(8,9) 2700	(9,10) 1000
(9,11) 1000	(9,12) 500	(9,13) 400	(10,14) 450	(11,14) 2900
(12,14) 2250	(13,14) 5150	(14,16) 30200	(15,16) 34500	(16,17) 34500
(17,18) 26200	(18,19) 26200	(19,20) 26200	(19,21) 26200	(21,22) 26200
(22,23) 26200	(23,24) 26200	(24,25) 34500	(25,26) 34500	(25,26) 34500
(26,27) 26200	(27,28) 11300	(28,29) 26200	(29,30) 26200	(30,31) 25700
(31,32) 34500	(32,33) 26200	(33,34) 34500	(34,35) 26200	(35,36) 14400
(36,37) 14400	(37,38) 28100	(38,39) 22000	(39,40) 32300	(40,41) 34500
(41,42) 34500	(42,43) 34500	(43,44) 34500	(44,45) 4900	(45,46) 34500
(46,47) 7000	(47,48) 34500	(48,49) 7000	(49,50) 26200	(50,51) 34500
(51,52) 34500	(52,sw) 34500			

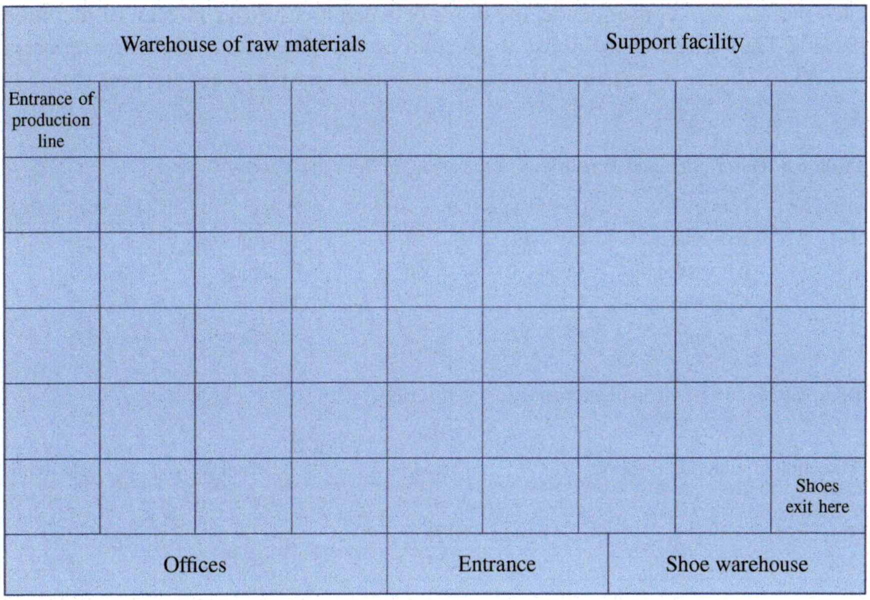

Fig. 6.1 Possible shoe manufacturer layout

"the flow from workstation 23 to workstation 24 is 26200". The abbreviation wrm stands for the warehouse of raw materials, and sw is the shoe warehouse.

This problem can be modelled as a QAP if we assume that every workstation has the same dimensions and that a layout of locations is fixed such that any workstation can be assigned to any location. One possible layout is shown in Fig. 6.1, where the entrance and exit of the production line are fixed (each being next to one of the two warehouses).

Alternatively, the problem can be modelled as an MRFLP or a UA-FLP if we make different assumptions on the dimensions of the workstations and the possible layout structures.

We have adapted the original application (see Sect. 6.10), which we observe is a dynamic layout problem.

6.5 Automotive Starter Battery Production

This example concerns a manufacturing plant that produces three types of automotive starter batteries. The plant is $24 \times 22 \,\mathrm{m}^2$, and it is divided into 8 areas (departments). The names and dimensions of the departments are presented in Table 6.8.

The specific problem is a practical application of dynamic facility layout. The production is divided into three time periods, and the production plan is different for each period. As a consequence, the flows between areas differ in each of the three periods. They are given in Table 6.9. The unit costs of the material flows are reported in Table 6.10, and it costs $500 to change one area's activity between periods.

Table 6.8 Description and dimensions of the areas for the battery factory

Area No.	Name	Dimensions (m)	Area No.	Name	Dimensions (m)
1	Raw materials	4.0×4.0	5	Tin soldering	3.2×2.5
2	Clip assembly	3.6×2.0	6	Punching	3.8×3.6
3	Coil wiring	3.0×2.5	7	Testing	4.0×3.5
4	Clamping	4.8×4.0	8	Packaging	4.0×3.6

Table 6.9 Flows between departments for each period

Direction of material transport	Quantity transported		
	Period 1	Period 2	Period 3
1-3	4300	1600	200
1-4	1000	0	5800
2-4	2400	5500	500
3-4	4300	1600	200
3-5	0	0	700
3-7	3400	7100	6300
4-5	7700	7100	6500
4-6	4300	1600	200
5-4	3400	5500	6300
5-6	4300	1600	200
6-3	3400	5500	6300
6-7	4300	1600	200
7-8	7700	7100	6500

Table 6.10 Flow costs between departments in the battery factory

Area	1	2	3	4	5	6	7	8
1	–	0.5	0.2	0.3	0.3	0.5	0.3	0.4
2		–	0.6	0.5	0.5	0.4	0.3	0.4
3			–	0.4	0.3	0.2	0.6	0.2
4				–	0.5	0.4	0.2	0.3
5					–	0.5	0.5	0.6
6						–	0.5	0.3
7							–	0.5
8								–

6.6 Building Layout with Evacuation Capacity

In this example, we want to assign departments to a seven-floor building with two sets of stairwells, minimizing the total flow. We must also assign each department to exactly one of the two stairwells so that, in the event of evacuation, we do not exceed the capacity of either stairwell (see Sect. 5.3.3).

We must assign 10 departments to 7 floors. The area of each floor is 15 square units, and Table 6.11 gives the areas of the departments. The flows between each pair of departments are symmetric, and they are as follows:

$$
\begin{pmatrix}
- & 10 & 0 & 3 & 0 & 5 & 4 & 5 & 5 & 2 \\
  & - & 5 & 4 & 5 & 0 & 0 & 0 & 4 & 4 \\
  &   & - & 1 & 0 & 3 & 4 & 7 & 4 & 7 \\
  &   &   & - & 4 & 6 & 10 & 3 & 0 & 2 \\
  &   &   &   & - & 5 & 3 & 8 & 7 & 2 \\
  &   &   &   &   & - & 7 & 8 & 3 & 9 \\
  &   &   &   &   &   & - & 9 & 1 & 0 \\
  &   &   &   &   &   &   & - & 3 & 3 \\
  &   &   &   &   &   &   &   & - & 9 \\
  &   &   &   &   &   &   &   &   & -
\end{pmatrix}
$$

In the event of evacuation, the average number of occupants going from each department to a stairwell is given in Table 6.12, and the floor-to-exit distances (via either stairwell) are listed in Table 6.13.

Table 6.11 Area of each department

Department	1	2	3	4	5	6	7	8	9	10
Area (squared units)	15	7.5	15	7.5	15	7.5	7.5	15	7.5	15

Table 6.12 Evacuation flows

Department	1	2	3	4	5	6	7	8	9	10
Evacuation flow (people/minute)	100	50	40	120	100	65	70	150	50	70

Table 6.13 Floor-to-exit distances

Floor	1	2	3	4	5	6	7
Distance	0	1	2	3	4	5	6

Table 6.14 Workstation dimensions for automotive example

Workstation	Dimensions ($m \times m$)
1, 2, 3, 4, 6, 8, 10, 11, 13	1×2
5, 14	2×2
7, 9	1×1
12	1×3

Table 6.15 Directions and flows for automotive example

Flows (direction and amount)				
(1,2) 5	(1,5) 1	(1,6) 5	(2,3) 5	(3,4) 5
(4,5) 5	(5,6) 6	(6,12) 11	(7,8) 4	(8,9) 4
(9,10) 4	(10,11) 4	(11,12) 4	(12,13) 15	(13,14) 15

6.7   Assembly of Automotive Components

This example comes from a company that assembles automotive components. It has 14 workstations, the dimensions of which are listed in Table 6.14. The flows between pairs of machines are presented in Table 6.15; each entry gives the pair of workstations and the flow between them.

This example can be modelled as an MRFLP or a UA-FLP by making different assumptions on the structure of the desired layout.

6.8   Production of Plastic Parts for the Automotive Industry

This example concerns the production of plastic parts for the automotive industry. The company has 3 facilities, geographically separated, and 13 possible locations for 13 departments, as illustrated in Fig. 6.2. When the possible positions for the departments are known in advance, the problem can be modelled as a variant of the QAP (see Sect. 5.1). This is different from multi-floor layout with three floors because there is no requirement for a common area on every floor to locate elevators and/or stairs.

The (symmetric) distances between the three facilities are given in the following matrix:

$$\begin{pmatrix} - & 30 & 50 \\ & - & 30 \\ & & - \end{pmatrix}.$$

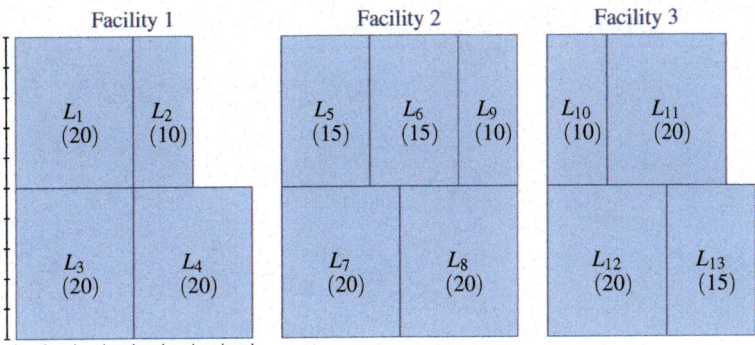

Fig. 6.2 Available locations for the 13 departments of the automotive parts example, with the area of each location given in brackets

Table 6.16 Transportation data for automotive parts example

	Period 1	Period 2	Period 3
Transportation costs inside the facilities	5	5	5
Transportation costs between facilities	8	10	15
Transportation capacity between facilities	1500	1800	2000

Table 6.17 Characteristics of the department types

Department type	Departments	Area			Rearrangement cost
		Period 1	Period 2	Period 3	
Raw material warehouse	Wr1	15	20	15	5000
	Wr2	20	20	20	
	Wr3	20	20	20	
Injection	I1	10	10	10	20000
	I2	10	10	10	
	I3	20	15	20	
Assembly	A1	20	20	20	10000
	A2	15	15	15	
	A3	15	15	15	
Painting	P	20	20	20	80000
Product warehouse	Wp1	20	20	20	5000
	Wp2	10	10	20	
	Wp3	20	20	20	

Table 6.16 gives the transportation cost inside each facility, the costs between facilities, and the transportation capacities between facilities, for three time periods. This example requires a dynamic version of the model for a fixed period.

The 13 departments can be aggregated into 5 department types, and their characteristics are given in Table 6.17. For each of the three time periods, the directed flows between departments are shown in Fig. 6.3.

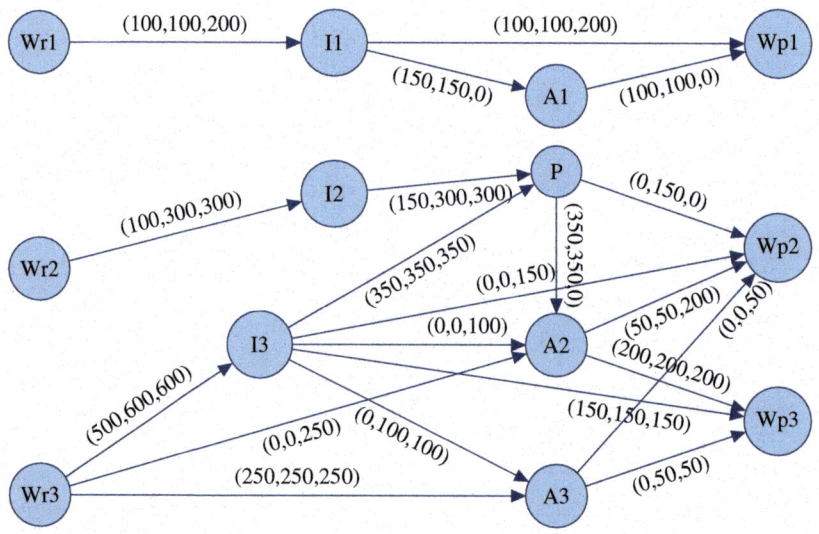

Fig. 6.3 Directed flows for each period

6.9   Applications to Hospital Layout

In this section we present three applications of facility layout in the context of hospitals. Section 6.9.1 presents a layout problem arising in the design of a new hospital, Sect. 6.9.2 concerns optimizing the room arrangement in an emergency service, and Sect. 6.9.3 presents a re-layout problem to reduce congestion in the patient flows.

6.9.1   Hospital Layout Design

This problem arose as part of the design of a German university hospital. The facility was required to have 2 floors and 30 departments. The most attractive grids are 3 × 5 or 4 × 4 cells for each floor. The latter grid has two more cells than departments. Table 6.18 gives the distance between every pair of departments, and Table 6.19 gives the pairwise flows. Because we know the distances between the cells in which to place the departments, the appropriate formulation for this application is the QAP (see Sect. 5.1).

Table 6.18 Distances between pairs of locations for the university hospital

	1	2	3	4	5	6	7	8	9	10	11	12	13	14	15	16	17	18	19	20	21	22	23	24	25	26	27	28	29	30
1	–	50	100	100	50	100	150	150	100	150	200	200	150	200	165	115	165	215	215	165	215	265	265	215	265	315	315	265	315	315
2		–	50	150	100	50	100	200	150	100	150	250	200	150	215	165	115	165	265	215	165	215	315	265	215	265	365	315	265	365
3			–	200	150	100	50	250	200	100	100	300	250	200	265	215	165	115	315	265	215	165	365	315	265	215	415	365	315	315
4				–	50	100	150	50	100	150	200	100	150	200	165	215	265	315	115	165	215	265	165	215	265	315	215	265	315	265
5					–	50	100	50	50	100	150	150	100	150	215	215	215	265	165	115	165	215	215	165	215	265	265	215	265	315
6						–	50	150	100	100	100	200	150	100	265	215	165	215	265	215	165	165	265	215	165	165	315	265	215	215
7							–	200	150	100	50	250	200	150	315	265	215	165	315	265	215	115	315	265	215	315	365	315	265	315
8								–	50	50	150	50	100	100	215	265	315	365	165	215	265	315	165	215	265	315	215	265	315	315
9									–	50	100	100	50	50	315	215	265	315	215	165	215	265	215	165	165	215	265	215	165	215
10										–	50	150	50	100	315	315	315	315	315	165	165	215	265	215	165	165	165	215	215	215
11											–	200	50	50	315	265	265	265	265	165	265	165	265	115	115	115	115	115	165	165
12												–	50	100	365	315	315	215	315	265	165	215	265	215	165	315	265	215	215	265
13													–	50	315	265	265	365	215	215	315	315	165	165	265	265	315	265	265	215
14														–	365	315	315	315	315	265	215	265	265	215	165	215	215	165	115	165
15															–	50	100	150	50	100	150	200	100	150	200	250	150	200	250	300
16																–	50	100	100	50	100	150	150	100	150	200	200	150	200	250
17																	–	50	150	100	50	100	200	150	100	150	250	200	150	200
18																		–	200	150	100	50	250	200	100	100	300	250	200	150
19																			–	50	100	150	50	100	150	200	100	150	200	250
20																				–	50	100	100	50	100	150	150	100	150	200
21																					–	50	150	100	50	100	200	150	100	150
22																						–	200	150	100	50	250	200	150	100
23																							–	50	100	150	50	100	150	200
24																								–	50	100	150	100	150	150
25																									–	50	50	100	50	100
26																										–	200	150	100	50
27																											–	50	100	150
28																												–	50	100
29																													–	50
30																														–

Table 6.19 Pairwise flows between departments

	1	2	3	4	5	6	7	8	9	10	11	12	13	14	15	16	17	18	19	20	21	22	23	24	25	26	27	28	29	30
1	—	2	2	2	2	2	2	2	3	0	3	2	0	0	0	3	3	0	0	2	1	2	1	1	0	3	0	0	0	0
2		—	2	2	3	2	3	2	0	0	2	1	0	0	2	3	3	0	0	2	1	2	1	1	0	3	0	0	0	0
3			—	2	2	3	0	0	0	0	2	2	0	0	2	3	3	0	0	2	1	2	1	1	0	3	0	0	0	0
4				—	2	2	3	1	0	0	2	1	0	0	4	3	3	0	0	2	1	2	1	1	0	3	0	0	0	0
5					—	2	3	2	0	0	2	1	0	0	2	3	3	0	0	2	1	2	1	1	0	3	0	0	0	0
6						—	0	0	0	0	2	2	0	0	2	3	3	0	0	2	1	2	1	1	0	3	0	0	0	0
7							—	4	4	3	3	0	4	0	0	2	2	0	0	3	4	1	0	0	0	0	0	0	0	0
8								—	0	0	0	2	0	0	0	3	3	0	0	3	1	1	0	0	0	0	0	0	0	0
9									—	0	0	1	4	0	0	2	2	0	0	2	0	1	0	0	0	0	0	4	0	0
10										—	0	0	0	0	0	0	0	0	0	0	0	1	1	0	0	0	0	0	0	0
11											—	3	4	0	0	3	3	0	0	0	1	2	0	0	0	2	0	4	0	0
12												—	3	0	0	1	1	0	0	0	0	2	2	0	0	0	0	0	0	0
13													—	4	0	3	3	0	0	2	1	1	1	0	0	0	0	4	0	0
14														—	0	0	0	4	0	0	0	1	0	0	0	3	0	0	0	0
15															—	1	1	0	0	0	0	2	0	0	0	3	0	2	0	0
16																—	4	2	0	0	0	1	0	0	0	2	0	0	0	0
17																	—	2	0	0	0	1	0	0	2	2	4	0	0	0
18																		—	0	0	0	1	0	0	2	4	4	0	0	0
19																			—	0	0	0	0	0	0	0	0	0	2	0
20																				—	1	0	0	0	0	0	0	0	3	0
21																					—	1	2	0	0	0	0	0	4	0
22																						—	1	0	0	0	0	0	0	0
23																							—	0	0	0	0	0	4	0
24																								—	2	0	0	0	4	0
25																									—	4	0	0	0	0
26																										—	0	0	0	4
27																											—	0	0	0
28																												—	0	0
29																													—	0
30																														—

Table 6.20 Room functions and dimensions for hospital emergency service

Number	Function of room	Width (m)	Length (m)
1	Emergency disposal	5.5	4.5
2	Gynaecology	5.5	4.5
3	Ophthalmic room	2.75	4.5
4	Otorhinolaryngology room	4	4.5
5	Dental room	2.4	4.5
6	Neurology room	3.6	4.5
7	Skin room	3.6	4.5
8	Surgical room	3.6	4.5
9	Orthopaedic room	2.8	4.5
10	Clinical laboratory	5.5	4.5
11	Electrocardiogram (ECG)	2.8	4.5
12	Blood tests	4.5	4.5
13	Paediatric room	4.5	4.5
14	Operating room	11	5.4
15	Medical emergency	9.1	1.5
16	Waiting room (registration)	9.1	3
17	Medicine bureau	12	8
18	Emergency observation	12	4.5
19	Emergency transfusion	11	4.5
20	Radiology	13.6	6.4

6.9.2   Emergency Service Layout

This example is concerned with the arrangement of rooms for an emergency service in a hospital. The dimensions of each room are in Table 6.20, and the flows of patients between rooms are in Table 6.21. In addition, there are closeness ratings between the departments (Table 6.22).

6.9.3   Rearrangement of a Hospital

This example involves the re-layout of a hospital located in Egypt. The hospital has six facilities: Out-patient, In-patient, Dental Research, Accident and Emergency, Physiotherapy and Housekeeping, and Maintenance. They will be located in separate buildings. The patients in the Out-patient Facility must move among its 17 clinics, and the locations of these clinics relative to each other was causing too much travelling for the patients, introducing bottlenecks and serious delays. This led to a study of the layout of the facility, the result of which was a reduction in the total distance travelled and hence in the frequency of bottlenecks.

Table 6.21 Flow of patients for hospital emergency service

Dep.	1	2	3	4	5	6	7	8	9	10	11	12	13	14	15	16	17	18	19	20
1	–	287	223	688	961	5684	229	10908	5797	0	0	0	2133	0	27751	0	0	1878	0	0
2		–	0	0	0	0	0	0	0	958	170	4	0	0	0	954	0	0	20	18
3			–	0	0	0	0	0	0	184	0	0	0	2	0	3177	0	0	44	632
4				–	0	0	0	0	0	1360	116	0	0	16	0	7882	20	0	1878	1298
5					–	0	0	0	1	262	12	0	0	10	0	2732	0	0	102	606
6						–	0	0	0	0	6122	2	0	0	0	23782	2	4	3792	9162
7							–	0	0	0	82	2	0	0	0	1882	0	0	1550	12
8								–	8	29186	2416	0	0	7	0	42851	26	4	0	15380
9									–	2298	554	0	0	246	0	18938	42	0	544	17162
10										–	0	0	1938	0	57678	150	0	1992	0	0
11											–	0	224	0	7613	0	0	332	0	0
12												–	0	0	0	1	0	0	0	0
13													–	0	0	5934	0	0	0	726
14														–	0	1	0	0	0	0
15															–	129011	0	2	53150	16314
16																–	0	4904	0	0
17																	–	0	0	0
18																		–	326	222
19																			–	0
20																				–

Table 6.22 Closeness rating between departments for hospital emergency service, ranging from 0 (proximity undesirable) to 5 (proximity absolutely necessary)

Department	1	2	3	4	5	6	7	8	9	10	11	12	13	14	15	16	17	18	19	20
1	–	4	4	4	4	4	4	4	4	4	4	4	4	4	4	4	3	3	3	3
2		–	2	2	2	2	5	3	0	4	4	4	5	3	4	4	3	3	3	4
3			–	5	5	2	0	2	1	4	4	4	2	3	4	4	3	3	3	4
4				–	5	2	0	2	1	4	4	4	2	3	4	4	3	3	3	4
5					–	2	0	2	1	4	4	4	2	3	4	4	3	3	3	4
6						–	2	5	1	4	4	4	2	3	5	4	3	3	3	4
7							–	3	0	4	4	4	5	3	4	4	3	3	3	4
8								–	3	4	4	4	5	3	4	4	3	3	3	4
9									–	4	4	4	2	4	4	4	3	3	3	4
10										–	5	5	4	4	4	4	3	3	3	4
11											–	5	4	4	4	4	3	3	3	4
12												–	4	4	4	4	3	3	3	4
13													–	4	4	4	3	3	3	4
14														–	4	4	3	3	3	4
15															–	4	3	3	3	4
16																–	3	3	3	4
17																	–	3	3	4
18																		–	3	4
19																			–	4
20																				–

Table 6.23 Clinics in the out-patient facility of the Egyptian hospital

Clinic number	Name	Clinic number	Name
1	Check-in room	11	X-Ray
2	General practitioner	12	Orthopaedic
3	Pharmacy	13	Psychiatric
4	Gynaecological and Obstetric	14	Squint
5	Medicine	15	Minor operations
6	Paediatric	16	Minor operations
7	Surgery	17	Dental
8	Otorhinolaryngology	18	Dental surgery
9	Urology	19	Dental prosthetic

The out-patient facility is composed of a check-in room, a waiting room, and 17 clinics. All the rooms and clinics require roughly the same area, with the exception of the Minor Operation section that occupies nearly double the typical space. It was therefore split into two facilities that have to be located next to each other. This gives a total of 19 facilities, listed in Table 6.23. There is no data on the sizes of the departments, so we formulate this problem as a QAP. Estimates of the patient flows between services are available on an annual basis. The distances between

Table 6.24 Distances between pairs of locations

	1	2	3	4	5	6	7	8	9	10	11	12	13	14	15	16	17	18	19
1	–	12	36	28	52	44	110	126	94	63	130	102	65	98	132	132	126	120	126
2		–	24	75	82	75	108	70	124	86	93	106	58	124	161	161	70	64	70
3			–	47	71	47	110	73	126	71	95	110	46	127	163	163	73	67	73
4				–	42	34	148	111	160	52	94	148	49	117	104	109	111	105	111
5					–	42	125	136	102	22	73	125	32	94	130	130	136	130	136
6						–	148	111	162	52	96	148	49	117	152	152	111	105	111
7							–	46	46	136	47	30	108	51	79	79	46	47	41
8								–	69	141	63	46	119	68	121	121	27	24	36
9									–	102	34	45	84	23	80	80	69	64	51
10										–	64	118	29	95	131	131	141	135	141
11											–	47	56	54	94	94	63	46	24
12												–	100	51	89	89	46	40	36
13													–	77	113	113	119	113	119
14														–	79	79	68	62	51
15															–	10	113	107	119
16																–	113	107	119
17																	–	6	24
18																		–	12
19																			–

pairs of departments are given in Table 6.24, and the flows are given in Table 6.25. The entries in Table 6.25 were obtained by averaging the flow between each pair of clinics, thus generating a symmetric matrix. The flow between facilities 15 and 16 (the two components of Minor Operations) is set to 99999 to force them to be adjacent in the optimal solution.

The distances between the locations were measured by tracing the paths taken by patients as they moved from one location to another. If the movement involved a change in floors, the corresponding vertical distance was multiplied by a subjective factor of 3. After visiting all the necessary clinics, the patient must return to the first one visited for administrative purposes. This essentially involves retracing the outward path because all the clinics are in the same building and there is only one main corridor per floor. Thus, the distance matrix can also be taken to be symmetric, even for pairs of locations on different floors.

6.10   References and Further Reading

The metallurgical industry instance presented in Sect. 6.1 comes from the master's thesis of Patil (1963). The manufacturing example in Sect. 6.2 is from El-Baz (2004). This example does not contain real-world data; we include it because

Table 6.25 Flows between pairs of departments

	1	2	3	4	5	6	7	8	9	10	11	12	13	14	15	16	17	18	19
1	–	76687	0	415	545	819	135	1368	819	5630	0	3432	9082	1503	0	0	13732	1368	1783
2		–	40951	4118	5767	2055	1917	2746	1097	5712	0	0	0	268	0	1373	268	0	0
3			–	3848	2524	3213	2072	4225	566	0	0	404	9372	0	972	0	13538	1368	0
4				–	256	0	0	0	0	829	128	0	0	0	0	0	0	0	0
5					–	0	0	0	47	1655	287	0	42	0	0	0	226	0	0
6						–	0	0	0	926	161	0	0	0	0	0	0	0	0
7							–	0	196	1538	196	0	0	0	0	0	0	0	0
8								–	0	0	301	0	0	0	0	0	0	0	0
9									–	1954	418	0	0	0	0	0	0	0	0
10										–	0	282	0	0	0	0	0	0	0
11											–	1686	0	0	0	0	0	0	0
12												–	0	0	0	0	0	0	0
13													–	0	0	0	0	0	0
14														–	0	0	0	0	0
15															–	99999	0	0	0
16																–	226	0	0
17																	–	0	0
18																		–	0
19																			–

the reader can compute the flows using the information provided. The defence application in Sect. 6.3 is from Balamurugan (2012).

The shoe manufacturing application in Sect. 6.4 is from Ulutas and Islier (2015), where it is considered in the broader context of the DFLP. We have presented the data for one period, i.e., we consider a static (as opposed to dynamic) layout problem. We thank the authors for providing this data.

The battery example in Sect. 6.5 is from Liu et al (2017); note that the matrix of flow costs in their Appendix A.2 is not symmetric, but this appears to be a typo. The evacuation example in Sect. 6.6 is from Hahn et al (2010), and the automotive assembly example in Sect. 6.7 is from Kovács (2020).

The plastic parts example in Sect. 6.8 is from Azevedo et al (2017). The authors propose a multi-objective approach for a QAP; their work includes additional technical details and data that are not included here.

In Sect. 6.9 three hospital applications are presented. The university hospital example in Sect. 6.9.1 is from Krarup and Pruzan (1978), and the data for this instance is available on the QAPLIB (Burkard et al 1997) accessible at https://www.miguelanjos.com/qaplib. The emergency service case in Sect. 6.9.2 is from Zuo et al (2019). Their matrix of patient flows includes some nonzero diagonal entries. These values arise when a patient visits the same department twice, but we have omitted them because in these cases the distance travelled by the patient is zero. They considered closeness ratings in addition to the flows; the ratings were introduced in Heragu (2008). Finally, the Egyptian hospital example in Sect. 6.9.3 is from Elshafei (1977). The data for this instance can also be found on the QAPLIB.

References

Azevedo MM, Crispim JA, Pinho de Sousa J (2017) A dynamic multi-objective approach for the reconfigurable multi-facility layout problem. J Manuf Syst 42:140–152

Balamurugan K (2012) Application of simulation and genetic algorithm for machine layout design. J Inf Optim Sci 33(6):653–664

Burkard RE, Karisch SE, Rendl F (1997) QAPLIB—a quadratic assignment problem library. J Global Optim 10(4):391–403

El-Baz MA (2004) A genetic algorithm for facility layout problems of different manufacturing environments. Comput Ind Eng 47(2):233–246

Elshafei AN (1977) Hospital layout as a quadratic assignment problem. Oper Res Q 28(1):167–179

Hahn P, MacGregor Smith J, Zhu YR (2010) The multi-story space assignment problem. Ann Oper Res 179(1):77–103

Heragu SS (2008) Facilities design. CRC Press

Kovács G (2020) Combination of lean value-oriented conception and facility layout design for even more significant efficiency improvement and cost reduction. Int J Prod Res 58(10):2916–2936

Krarup J, Pruzan PM (1978) Computer-aided layout design. Springer, Berlin, Heidelberg, pp 75–94

Liu J, Wang D, He K, Xue Y (2017) Combining Wang–Landau sampling algorithm and heuristics for solving the unequal-area dynamic facility layout problem. Eur J Oper Res 262(3):1052–1063

Patil NF (1963) Quantitative techniques for plant layout analysis. Master's thesis, Kansas State University, Manhattan, Kansas, USA

Ulutas B, Islier AA (2015) Dynamic facility layout problem in footwear industry. J Manuf Syst 36:55–61

Zuo X, Li B, Huang X, Zhou M, Cheng C, Zhao X, Liu Z (2019) Optimizing hospital emergency department layout via multiobjective tabu search. IEEE Trans Autom Sci Engineering 16(3):1137–1147

Appendix A
Semidefinite Optimization and Conic Optimization

Semidefinite optimization (SDO) is the class of optimization problems in which we seek to maximize or minimize a linear function of the elements of a matrix variable subject to linear constraints on those elements, together with a constraint that the entire matrix must be symmetric positive semidefinite (PSD). The PSD condition is equivalent to requiring that all the eigenvalues of the matrix must be greater than or equal to zero. LO is a special case of SDO that corresponds to the matrix variable being a diagonal matrix, i.e., a matrix with all the elements off the main diagonal equal to zero. Another special case of SDO is second-order conic optimization (SOCO), which corresponds to optimizing over the second-order cone (see Sect. A.2).

SDO problems are important because they are solvable in polynomial time (see Sect. 2.1.1), so any problem that can be expressed using SDO is also solvable in polynomial time. Moreover, SDO problems can be solved efficiently in practice, by using an available software package or implementing a specialized algorithm.

SDO has a number of similarities with LO. Like LO problems, SDO problems come in pairs. One of the problems is referred to as the *primal*, and the other is the *dual*. Either problem can be chosen as the primal, since the two problems are dual to each other. The most common standard formulation of SDO is as follows:

$$\text{(P) } \min \langle C, X \rangle \qquad\qquad \text{(D) } \max b^T y$$
$$\text{s.t. } \langle A_i, X \rangle = b_i, i = 1, \ldots, m \qquad \text{s.t. } \sum_{i=1}^{m} y_i A_i + S = C \qquad \text{(A.1)}$$
$$X \succeq 0 \qquad\qquad\qquad\qquad S \succeq 0$$

where (P) denotes the primal and (D) the dual. The variables X and S are in $\mathbb{S}^n$, the space of $n \times n$ real symmetric matrices; $X \succeq 0$ indicates that the matrix X is positive semidefinite; the data matrices $A_i \in \mathbb{S}^n$ and $C \in \mathbb{S}^n$ may be assumed without loss

© Springer Nature Switzerland AG 2021
M. F. Anjos, M. V. C. Vieira, *Facility Layout*, EURO Advanced Tutorials
on Operational Research, https://doi.org/10.1007/978-3-030-70990-7

of generality to be symmetric; and $b \in \mathbb{R}^m$ and $y \in \mathbb{R}^m$ are column vectors. We use the inner product between two matrices in $\mathbb{S}^n$ defined as

$$\langle R, S \rangle := \text{trace}(RS) = \sum_{i=1}^{n} \sum_{j=1}^{n} R_{ij} S_{ij}$$

where trace (M) denotes the trace of the square matrix M, which is the sum of its diagonal elements. It is usually assumed, without loss of generality, that the matrices $A_i, i = 1, \ldots, m$, are linearly independent.

The dual SDO problem in (A.1) can equivalently be written without using the dual variable S:

$$\max b^T y$$
$$\text{s.t.} \quad C - \sum_{i=1}^{m} y_i A_i \succeq 0,$$

where the inequality constraint is interpreted as follows:

$$A - B \succeq 0 \quad \Leftrightarrow \quad A \succeq B.$$

A.1   Positive Semidefinite Matrices

Positive semidefinite matrices have numerous properties. For a 2×2 symmetric matrix, the necessary and sufficient conditions for positive semidefiniteness are

$$\begin{pmatrix} x_{11} & x_{12} \\ x_{12} & x_{22} \end{pmatrix} \succeq 0 \Leftrightarrow x_{11} \geq 0, \ x_{22} \geq 0, \ \text{and} \ x_{11}x_{22} - x_{12}^2 \geq 0. \tag{A.2}$$

This is a special case of Theorem A.1 below. To state the theorem, we need the following definition.

Definition A.1 If $X \in \mathbb{S}^n$, then for every nonempty subset $I \subseteq \{1, 2, \ldots, n\}$, the *principal submatrix* of X corresponding to I, denoted by $X(I)$, is the square submatrix with rows and columns indexed by I. The determinant of $X(I)$ is called the *principal minor* of X corresponding to I.

Theorem A.1 *For $X \in \mathbb{S}^n$, X is PSD if and only if all the principal minors of X are non-negative.*

Example A.1 An important example of an SDO problem is

$$\min \langle C, X \rangle$$
$$\text{s.t.} \quad x_{ii} = 1, \quad i = 1, \ldots, n \tag{A.3}$$
$$X \succeq 0.$$

This SDO problem is particularly relevant in facility layout. Its feasible set is the set of all symmetric matrices $n \times n$ that are PSD and have ones on the diagonal. Note that Theorem A.1 implies that $x_{ij}^2 \leq 1$ holds for every off-diagonal element of X feasible for (A.3). To see why this is true, consider the 2×2 principal submatrix $X(\{1, 2\})$ and apply (A.2):

$$\begin{pmatrix} x_{ii} & x_{ij} \\ x_{ij} & x_{jj} \end{pmatrix} \geq 0 \quad \Rightarrow \quad x_{ii} x_{jj} - x_{ij}^2 \geq 0 \quad \Leftrightarrow \quad x_{ij}^2 \leq 1$$

because $x_{ii} = 1$ and $x_{jj} = 1$.

A useful characterization of PSD matrices is the following:

Theorem A.2 *An $n \times n$ symmetric matrix X is PSD if and only if $y^T X y \geq 0$ for all vectors $y \in \Re^n$.*

We can use this theorem to show that the matrices $\Gamma = g\,g^T$ defined by equation (2.61) in Sect. 2.7 are PSD for any choice of g. This is because for any vector y:

$$y^T \Gamma y = y^T g\,g^T y = (g^T y)^2 \geq 0.$$

A.2 Second-Order Conic Optimization

The $(n + 1)$-dimensional second-order cone (SOC) is defined as

$$\mathrm{SOC}^{n+1} = \left\{ (x_0, x_1, \ldots, x_n) \in \mathbb{R}^{n+1} \mid x_0 \geq \sqrt{x_1^2 + \ldots + x_n^2} \right\}.$$

An equivalent expression can be given in the form of a rotated quadratic cone:

$$\mathrm{SOC}^{n+1} = \left\{ (x_0, x_1, \ldots, x_n) \in \mathbb{R}^{n+1} \mid 2x_0 x_1 \geq x_2^2 + \ldots + x_n^2, x_0 \geq 0, x_1 \geq 0 \right\}.$$

Mathematically, the latter is a rotation of the former through an angle of 45 degrees in the (x_0, x_1)-plane, and for modelling purposes the rotated form is often more convenient.

A non-negativity constraint $x_0 \geq 0$ is just a SOC constraint in a space of dimension 1 ($n = 0$); hence, LO is a special case of SOCO. For dimension 2, the SOC can be expressed as

$$\mathrm{SOC}^2 = \left\{ (x_0, x_1) \in \mathbb{R}^2 \mid x_0 \geq |x_1| \right\},$$

which is a rotated non-negative quadrant; hence, SOCO in dimension 2 is also a LO problem. For dimensions 3 and greater, the SOC is not polyhedral, and hence in general, SOCO is not equivalent to LO. For instance, it is straightforward to verify that a 2×2 PSD constraint is equivalent to a 3-dimensional SOC constraint:

$$\begin{pmatrix} x_{11} & x_{12} \\ x_{12} & x_{22} \end{pmatrix} \geq 0 \Leftrightarrow x_{11}x_{22} \geq x_{12}^2 \text{ and } x_{11}, x_{22} \geq 0 \Leftrightarrow \begin{pmatrix} x_{11} + x_{22} \\ x_{11} - x_{22} \\ 2x_{12} \end{pmatrix} \in \text{SOC}^3.$$

$$(A.4)$$

This result can be proved using basic algebra, and it is left as an exercise for the reader.

The SOCO primal–dual pair has the form:

$$(P_{\text{SOC}}) \inf \sum_{j=1}^{\ell} c_j^T x_j \qquad\qquad (D_{\text{SOC}}) \sup b^T y$$

$$\text{s.t. } \sum_{j=1}^{\ell} A_j x_j = b, \qquad\qquad \text{s.t. } A_j^T y + s_j = c_j, j = 1, \ldots, \ell$$

$$x_j \in \text{SOC}^{n_j+1}, j = 1, \ldots, \ell \qquad\qquad s_j \in \text{SOC}^{n_j+1}, j = 1, \ldots, \ell$$

$$(A.5)$$

where ℓ is the number of SOCs.

SOCO is a special case of SDO in the sense that

$$(x_0, x_1, \ldots, x_n) \in \text{SOC}^{n+1} \Leftrightarrow \begin{pmatrix} x_0 & 0 & 0 & 0 & x_1 \\ 0 & x_0 & 0 & 0 & x_2 \\ 0 & 0 & \ddots & 0 & \vdots \\ 0 & 0 & 0 & x_0 & x_n \\ x_1 & x_2 & \cdots & x_n & x_0 \end{pmatrix} \geq 0. \qquad (A.6)$$

While (A.4) can be checked easily, this more general statement about SOCs requires more advanced matrix theory (see Sect. A.3).

The equivalence (A.6) makes it possible in principle to solve SOCO problems by converting them to SDO problems. However, this is much less efficient than making use of the SOC structure.

A.3   References and Further Reading

Conic optimization is a thriving research area, and this Appendix merely scratches the surface of this large and growing field. A more detailed introduction to SDO, including a discussion on the software available for SOCO and SDO problems, can be found in Anjos (2017). A wealth of results about PSD matrices can be found in Chapter 7 of Horn and Johnson (1990), including the statement and proof of

Theorem A.1. Theorem A.2 is stated in Horn and Johnson (1990, Section 7.1) as the definition of a PSD matrix (there are various equivalent ways of defining positive semidefiniteness).

SDO, often referred to as semidefinite programming or SDP for short, has been studied in different forms since at least the 1940s. The interest in SDO grew dramatically during the 1990s because of the extension of polynomial-time interior-point methods for LO to the solution of SDO problems, and hence to SOCO problems. Some applications of SDO followed soon after this development, such as the solution of linear matrix inequalities in control theory, and the design of polynomial-time approximation schemes for the maximum-cut and stable-set problems. This outburst of activity led to the publication of the *Handbook of Semidefinite Programming* (Wolkowicz et al 2000) that provided an account of much of the activity in SDO up to 2000.

The research activity continued into the 2000s and increased further via interactions with algebraic geometry through the close connections between semidefinite matrices and polynomial optimization problems. This decade of developments brought about several important new results and is documented in the *Handbook on Semidefinite, Conic and Polynomial Optimization* (Anjos and Lasserre 2011). In particular, these developments raised the profile and importance of polynomial optimization; see, e.g., Lasserre (2010). Much of the ongoing activity can be followed on preprint websites such as *ArXiv* (https://arxiv.org) and *Optimization Online* (https://www.optimization-online.org).

A detailed treatment of the important SDO problem (A.3) is given in Wolkowicz and Anjos (2002). The links between (A.3) and the SDO relaxation of the stable-set problem underlying Lovász's famous theta function were studied in Laurent et al (1997). The equivalence of a 2×2 PSD constraint and a 3-dimensional SOC constraint was first observed and applied in Kim and Kojima (2003). The equivalence (A.6) follows from the Schur complement theorem; this result is stated as Theorem 7.7.6 in Horn and Johnson (1990). Alizadeh and Goldfarb (2003) and Lobo et al (1998) provide in-depth presentations of the SOC, algorithms for SOCO, and classes of optimization problems that can be formulated using SOCO.

References

Alizadeh F, Goldfarb D (2003) Second-order cone programming. Mathematical Programming 95(1):3–51

Anjos MF (2017) Chapter 9: Conic linear optimization. In: Advances and trends in optimization with engineering applications. SIAM, pp 107–120

Anjos MF, Lasserre JB (2011) Handbook on semidefinite, conic and polynomial optimization, vol 166. Springer Science & Business Media

Horn R, Johnson C (1990) Matrix analysis. Cambridge University Press, Cambridge

Kim S, Kojima M (2003) Exact solutions of some nonconvex quadratic optimization problems via SDP and SOCP relaxations. Comput Optim Appl 26(2):143–154

Lasserre JB (2010) Moments, positive polynomials and their applications, vol 1. World Scientific

Laurent M, Poljak S, Rendl F (1997) Connections between semidefinite relaxations of the max-cut
    and stable set problems. Mathematical Programming 77(1):225–246
Lobo MS, Vandenberghe L, Boyd S, Lebret H (1998) Applications of second-order cone
    programming. Linear Algebra Appl 284(1–3):193–228
Wolkowicz H, Anjos MF (2002) Semidefinite programming for discrete optimization and matrix
    completion problems. Discrete Appl Math 123(1–2):513–577
Wolkowicz H, Saigal R, Vandenberghe L, Vandenberghe R (2000) Handbook of semidefinite
    programming: Theory, algorithms, and applications, vol 27. Springer Science & Business
    Media

Printed by Printforce, United Kingdom